The Ultimate UKCAT Guide

UniAdmissions

Published by *RAR Medical Services Limited*
www.uniadmissions.co.uk
info@uniadmissions.co.uk
Tel: 0203 375 6294

The Ultimate UKCAT Guide

1250 Practice Questions

Dr David Salt
Dr Rohan Agarwal

About the Authors

David is **Director of Services** at *UniAdmissions*, taking the lead in product development and customer service. David read medical sciences at Gonville and Caius College Cambridge, graduating in 2012, completed his clinical studies in the Cambridge Clinical School and now works as a doctor in Leicester.

David is an experienced tutor, having helped students with all aspects of the university applications process. He has authored five books to help with university applications and has edited four more. Away from work, David enjoys cycling, outdoor pursuits and good food.

Rohan is **Director of Operations** at *UniAdmissions* and is responsible for its technical and commercial arms. He graduated from Gonville and Caius College, Cambridge, and is now a doctor in Leicester. Over the last five years, he has tutored hundreds of successful Oxbridge and Medical applicants. He has authored ten books on admissions tests and interviews.

He has taught physiology to undergraduates and interviewed medical school applicants for Cambridge. He has also published research on bone physiology and writes education articles for the Independent and Huffington Post. In his spare time, Rohan enjoys playing the piano and table tennis.

The Basics

What is the UKCAT?

The United Kingdom Clinical Aptitude Test (UKCAT) is a two hour computer based exam that is taken by applicants to medical and dental schools. The questions are randomly selected from a huge question bank. Since every UKCAT test is unique, candidates can sit the UKCAT at different times. There is a three month testing period and you can sit the test anytime within it.

You register to sit the test online and book a time slot. On the day, bring along a printout of your test booking confirmation and arrive in good time. Your identity will be checked against a photographic ID that you'll need to bring. You then leave your personal belongings in a locker and enter the test room. Make sure you go to the toilet and have a drink before going in, to save wasting time during the test.

Test Structure:

SECTION	WHAT DOES IT TEST?	QUESTIONS	TIME
ONE	**Verbal Reasoning**: Essentially a reading comprehension test, with questions based on passages. This is a test of accuracy and speed of reading.	44	22 minutes
TWO	**Decision Making:** Tests your application of logic and approach to problem solving. Developing a structured approach will help you succeed.	29	32 minutes
THREE	**Quantitative Reasoning**: The maths is usually intellectually straightforward but can involve complex or multi stage calculations against the clock. Assesses your understanding of numbers.	36	25 minutes
FOUR	**Abstract Reasoning**: Matching shapes to which set they belong in. This tests your pattern recognition skills and rewards those who have clear and logical organisation of thought.	55	14 minutes
FIVE	**Situational Judgement**: This tests your ability to make decisions in a clinical environment. To perform well, you must understand the role and responsibilities of a medical student.	67	27 minutes

Who has to sit the UKCAT?

You have to sit UKCAT if you are applying for any of the universities that ask for it in the current application cycle. You are strongly advised to check this list in May to see if the universities you are considering require it. The following is a list of the universities and courses requiring the UKCAT for 2018 entry. As it is subject to change, it is included for guidance only:

University Name	Courses requiring UKCAT
University of Aberdeen	A100, A201
University of Birmingham	A100, A200
University of Bristol	A100, A108, A206, A208
Cardiff University	A100, A104, A200, A204
University of Dundee	A100, A104, A200, A204
University of East Anglia	A100, A104
University of Edinburgh	A100
University of Exeter	A100
University of Glasgow	A100, A200
Hull York Medical School	A100
Keele University	A100, A104
King's College London	A100, A101, A102, A202, A205, A206
University of Leicester	A100
University of Liverpool	A100, A200, A201
University of Manchester	A104, A106, A204, A206, B840
University of Newcastle	A100, A101, A206
University of Nottingham	A100, A108
Plymouth University	A100, A206
Queen Mary, University of London	A100, A101, A110, A200
Queen's University Belfast	A100, A200
University of Sheffield	A100, A200
University of Southampton	A100, A101, A102
University of St Andrews	A100, A990
St George's, University of London	A100, A900
University of Warwick	A101
Poznan University, Poland	Advanced MD programme for graduates
American University of the Caribbean	4 year MD programme

The following table gives the subject area for each of the course codes that require UKCAT.

Course Code		Course Details
	A100	Undergraduate medicine, full time
	A101	Graduate-entry medicine, full time
	A102	Graduate-entry medicine, full time
	A104	Undergraduate medicine with preliminary year, full time
	A200	Undergraduate dentistry, full time
	A202	Graduate-entry dentistry, full time
	A204	Undergraduate dentistry with preliminary year, full time
	A205	Undergraduate dentistry, full time
	A206	Undergraduate dentistry, full time
(St. George's)	A900	Undergraduate medicine for international, full time
(St. Andrew's)	A990	Undergraduate medicine Canadian programme, full time
	B750	Dental hygiene and therapy
	B752	Dental hygiene and therapy
	B840	Oral Health Science

Why is the UKCAT used?

For medical schools, choosing the best applicants is hard. Every year, medical schools are flooded with talented applicants, all of whom have top grades and great personal statements. They felt they needed extra information to help them find the very best from the pool of very talented applicants they always had. That's why admissions tests were created.

The UKCAT was first introduced in 2006 to help medical schools make their choices. The test examines skills in different areas, all of which are related to critical thinking and decision making. The idea was to create a pure aptitude test, something which cannot be prepared for. However this is certainly not the case with the UKCAT, and this is even acknowledged by UKCAT itself based on their research. In our experience, you can improve your UKCAT score with only a small amount of work, and with proper organised preparation the results can be fantastic.

When do I sit the UKCAT?

When you register to sit UKCAT online, you choose your date and time slot and also the test centre. The UKCAT can be sat from 3rd July through to early October. Registration for the test opens in early May – we recommend you book your test early so you have the best choice of possible dates.

Many students find it helpful to sit UKCAT in mid-late August – this gives you time in the summer to prepare, but gets the test complete before you go back to school, so you have one less thing to worry about at that busy time. Remember that **you may want to modify your university choices** based on your UKCAT score to maximise your chances of success.

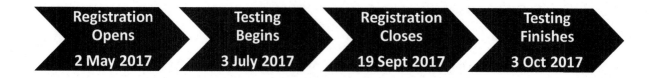

Registration Opens — 2 May 2017 | Testing Begins — 3 July 2017 | Registration Closes — 19 Sept 2017 | Testing Finishes — 3 Oct 2017

How much does it cost?

In the EU, tests sat between 3 July and 31 August cost £65. Tests sat between 1 September and 3 October cost £85. Tests outside the EU cost £115 throughout the testing period.

Some candidates who might struggle to fund the UKCAT fee are eligible for a bursary. If eligible to apply for one, you need to apply with supporting evidence by the deadline of 21 September 2017.

Bursary Eligibility Criteria

➢ Receipt of 16–19 bursary or EMA.
➢ Receipt of discretionary learner support.
➢ Receipt of full maintenance grant or special support grant.
➢ Receipt of income support, job seeker's allowance or employment support allowance.
➢ Receipt of universal credit.
➢ Receipt of child tax credit.
➢ Receipt of free school meals.
➢ Receipt of EU means tested support.
➢ Living with a family member in support of income support, job seeker's allowance or child tax credit.
➢ Awarded asylum support.

Can I re-sit the UKCAT?

You cannot re-sit UKCAT in the same application cycle – whatever score you get is with you for the year. That's why it's so important to make sure you're well prepared and ready to perform at your very best on test day.

If I reapply, do I have to resit the UKCAT?

If you choose to re-apply the following year, you need to sit UKCAT again. You take your new score with you for the new applications cycle. UKCAT scores are only valid for one year from the test date.

When do I get my results?

Because the test is computerised, results are generated immediately and you will be given your score on the day of the test. You will be given a printed sheet with your details and your score to take away. Knowing your score is useful as it can help you choose your universities tactically to maximise your chances of success. Note that you don't put your UKCAT score anywhere on your UCAS form, nor do you contact any universities to inform them. Universities that request UKCAT are sent your scores directly by UKCAT, so you don't need to do anything besides apply through UCAS.

Where do I sit the UKCAT?

UKCAT is a computerised exam and is sat at computer test centres, similar to the driving theory test. When you book the test, you choose the most convenient test centre to sit it at.

How is the UKCAT Scored?

When you finish the test, the computer works out your raw score by adding up your correct responses. There are no mark deductions for incomplete or incorrect answers, so it's a good idea to answer every question even if it's a guess. For the four cognitive sections, this is then scaled onto a scale from 300 – 900. The totals from each of these sections are added together to give your overall score out of 3,600. The new decision making section is now fully introduced into the test, and is scored as normal.

For section 5 (the situational judgement test, SJT) the scoring is slightly different. Here the appropriateness of your responses is used to generate a banding, from band 1 (being the best) to band 4 (being the worst). This is presented separately to the numerical score, such that every candidate's score contains a numerical score out of 3,600 and an SJT banding.

How does my score compare?

This is always a tough question to answer, but it makes sense to refer to the average scores. The scaling is such that around 600 represents the average score in any section, with the majority of candidates scoring between 500 and 700. Thus a score higher than 700 is very good and a score less than 500 is very weak.

For reference, in the 2017 entry cycle the mean scores at the end of testing were 573, 690 and 630 for sections 1, 3 and 4 respectively (section 2 was not marked, but this year it will be), giving an overall score of 1,893 or a mean score of 631 per section.

How is the UKCAT used?

Different universities use your UKCAT score in different ways. Firstly, universities that do not explicitly subscribe to UKCAT cannot see your UKCAT score and are unaware whether or not you sat UKCAT. Universities that use UKCAT can use it in a variety of ways – some universities use it as a major component of the assessment such as selecting candidates for interview based upon the score. Others use it as a smaller component, for example to settle tie-breaks between similar candidates. Each university publishes guidance on how they use the UKCAT, so you should check this out for the universities you are considering.

It's important to know how UKCAT is used in order to maximise your chances. If you score highly in UKCAT, you might decide to choose universities that select for interview based on a high UKCAT score cut-off. That way, you help to stack the odds in your favour – you might, for instance, convert a one in eight chance to a one in three chance. If your score isn't so good, consider choosing universities that don't use UKCAT in that way, otherwise you risk falling at the first hurdle and never getting the chance to show them how great you actually are.

By this logic, **it makes sense for all medical applicants to sit UKCAT** – if you score well it opens doors, and if you don't you don't even have to apply to UKCAT universities. It makes sense not to place all your eggs in one basket. If you were to, for example, apply to only BMAT universities, you risk jeopardising your entire application if you are unlucky on test day.

Can I qualify for extra time?

Yes – some people qualify for extra time in the UKCAT, sitting what is known as UKCAT SEN. If you usually have extra time in public exams at school, you are likely to be eligible to sit the UKCAT SEN. The overall time extension is 30 minutes, bringing the total test time up from 120 to 150 minutes; this is allocated proportionately across the different sections. If you have any medical condition or disability that may affect the test, requiring any special provision, or requiring you to take any medical equipment or medication into the test you should contact customer services to discuss how to best proceed.

General Advice

Practice

Preparing for the UKCAT will almost certainly improve your UKCAT score. You are unlikely to be familiar with the style of questions in sections 3, 4 and 5 of the UKCAT when you first encounter them. With practice, you'll become much quicker at interpreting the data and your speed will increase greatly. Practising questions will put you at ease and make you more comfortable with the exam format, and you will learn and hone techniques to improve your accuracy. This will make you calm and composed on test day, allowing you to perform at your best.

Initially, **work through the questions at your own pace**, and spend time carefully reading the questions and looking at the additional data. The purpose of this is to gain familiarity with the question styles and to start to learn good techniques for solving them. Then closer to test day, make sure you practice the questions under exam conditions and at the correct pace.

Start Early

It is much easier to prepare if you practice little and often. Start your preparation well in advance, ideally by early July but you are advised to start no later than early August. This way, you will have plenty of time to work through practice questions, to build up your speed and to incorporate time-saving techniques into your approach. How to start – well by reading this you're obviously on track!

How to Work

Although this obviously depends on your learning style, it can be helpful to split your preparation into two stages. Early on, it's often best to focus on only one section per day. Firstly read about the section, then maybe follow through a fully worked example, then try some practice questions, stop and mark them and work out anywhere you've gone wrong. By working on only one section per day, you focus your thoughts and allow yourself to get deeper into understanding the question type you're working on. The aim of early preparation is to *learn* about the test and the question styles. Don't worry so much about timing as you do about accuracy and technique. It is a good idea to start on the verbal reasoning and quantitative reasoning as these take the greatest amount of time to improve.

Nearer to test day, you'll need to work on multiple sections per day to help train your thinking to switch quickly from one mode to another. Start to attempt questions with strict timing. The aim now is that you're comfortable in answering the questions, so the next step is to work on exam technique to ensure you know what to expect on the day. Attempt the online UKCAT practice questions – although there are not very many of these, they are set out as the computer will be on the day with the official UKCAT calculator, so getting familiar with this is important.

Repeat Tough Questions

When checking through answers, pay particular attention to questions you have got wrong. Look closely through the worked answers in this book until you're confident you understand the reasoning- then repeat the question later without help to check you can now do it. If you use other resources where only the answer is given, have another look at the question and consider showing it to a friend or teacher for their opinion.

Statistics show that without consolidating and reviewing your mistakes, you're likely to make the same mistakes again. Don't be a statistic. Look back over your mistakes and address the cause to make sure you don't make similar mistakes when it comes to the test. You should avoid guessing in early practice. Highlight any questions you struggled with so you can go back and improve.

Positive Marking

When it comes to the test, the marking scheme is only positive – you won't lose points for wrong answers. You gain a mark for each correct answer and do not gain one for each wrong or unanswered one. Therefore if you aren't able to answer a question fully, you should guess. Since each question provides you with 3 to 5 possible answers, you have a 33% to 20% chance of guessing correctly – something which is likely to translate to a number of points across the test as a whole.

If you do need to guess, try to make it an educated one. By giving the question a moment's thought or making a basic estimation, you may be able to eliminate a couple of options, greatly increasing your chances of a successful guess. This is discussed more fully in the subsections.

Booking your Test

Unless there are strong reasons otherwise, you should try to **book your test during August**. This is because in the summer, you should have plenty of time to work on the UKCAT and not be tied up with schoolwork or your personal statement deadline. If you book it any earlier, you'll have less time for your all-important preparation; if you book any later, you might get distracted with schoolwork, your personal statement deadline and the rest of your UCAS application. In addition, you pay £20 more to sit the test late in the testing cycle.

Mock Papers

There are 2 full UKCAT papers freely available online at **www.ukcat.ac.uk** and once you've worked your way through the questions in this book, you are highly advised to attempt both of them and check your answers afterwards. There are also a further 2 full mock papers available at **www.uniadmissions.co.uk/ukcat-mock-papers**.

Prioritising Sections

Many students find sections 3 and 4 the easiest to improve on. Initially, section 4 can often be the hardest section, but because you start form a lower point with a bit of familiarity and practice your score can increase greatly. Likewise you can achieve good gains with section 3. Although the subject matter of the questions is not new, with practice you gain familiarity with the style of UKCAT questions and with using the calculator. This familiarity can give you a useful speed boost, increasing your score. If you start your preparation late, it would be wise to concentrate most on these sections in order to achieve the best gains.

A word on timing...

"If you had all day to do your UKCAT, you would get 100%, 3600 points. But you don't."

Whilst this isn't completely true, it illustrates a very important point. Once you've practiced and know how to answer the questions, the clock is your biggest enemy. This seemingly obvious statement has one very important consequence. **The way to improve your UKCAT score is to improve your speed.** There is no magic bullet. But there are a great number of techniques that, with practice, will give you significant time gains, allowing you to answer more questions and score more marks.

Timing is tight throughout the UKCAT – mastering timing is the first key to success. Some candidates choose to work as quickly as possible to save up time at the end to check back, but this is generally not the best way to do it. UKCAT questions have a lot of information in them – each time you start answering a question it takes time to get familiar with the instructions and information. By splitting the question into two sessions (the first run-through and the return-to-check) you double the amount of time you spend on familiarising yourself with the data, as you have to do it twice instead of only once. This costs valuable time. In addition, candidates who do check back may spend 2–3 minutes doing so and yet not make any actual changes. Whilst this can be reassuring, it is a false reassurance as it has no effect on your actual score. Therefore it is usually best to pace yourself very steadily, aiming to spend the same amount of time on each question and finish the final question in a section just as time runs out. This reduces the time spent on re-familiarising with questions and maximises the time spent on the first attempt, gaining more marks.

There is an option to flag questions for review, making it easier to check back if you have time at the end of the section. There is absolutely no disadvantage to using this. If you've guessed a question then it makes sense to mark it, so you know where to best spend your spare time if you do finish the section early. Always select an answer first time round (even if it's a guess), as there is no negative marking – you may not have time to turn back later on.

It is essential that you don't get stuck with the hardest questions – no doubt there will be some. In the time spent answering only one of these you may miss out on answering three easier questions. If a question is taking too long, choose a sensible answer and move on. Never see this as giving up or in any way failing, rather it is the smart way to approach a test with a tight time limit. With practice and discipline, you can get very good at this and learn to maximise your efficiency. It is not about being a hero and aiming for full marks – this is essentially impossible and in any case completely unnecessary. It is about maximising your efficiency and gaining the maximum possible number of marks within the time you have.

Verbal Reasoning

The Basics

Section 1 of the UKCAT is the verbal reasoning subtest. It tests your ability to quickly read a passage, find information that is relevant and then analyse statements related to the passage. There are 44 questions to answer and in 21 minutes, so you have just under 30 seconds per question. As with all UKCAT sections, you have one minute to read the instructions. The idea is that this tests both your language ability and your ability to make decisions, traits which are important in a good doctor.

You are presented with a passage, upon which you answer questions. Typically, there are 11 separate passages, each with 4 questions about it. There are two styles of question in section 1, and each requires a slightly different approach. All questions start with a statement relating to something in the passage.

In the first type of question, you are asked if the statement is true or false based on the passage. There is also the option to answer "cannot tell". Choose "true" if the statement either matches the passage or can be directly inferred from it. Choose "false" if the statement either contradicts the passage or exaggerates a claim the passage makes to an extent that it becomes untrue.

Choosing the "cannot tell" option can be harder. Remember that you are answering based *ONLY* on the passage and not on any of your own knowledge – so you choose the "cannot tell" option if there is not enough information to make up your mind one way or the other. Try to choose this option actively. "Cannot tell" isn't something to conclude too quickly, it can often be the hardest answer to select. Choose it when you're actively looking for a certain piece of information to help you answer a question, and you cannot find it.

In the other type of question, you are given a stem and have to select the most appropriate response based on the question. There is only one right answer – if more than one answer seems appropriate, the task is to choose the *best* response. Remember that there is no negative marking in the UKCAT. There will be questions where you aren't certain. If that is the case, then choose an option that seems sensible to you and move on. A clear thought process is key to doing well in section 1 – you will have the opportunity to build that up through the worked examples and practice questions until you're answering like a pro!

This is the first section of the UKCAT, so you're bound to have some nerves. Ensure that you have been to the toilet because once the exam starts you can't pause and go. Take a few deep breaths and calm yourself down. Try to shut out distractions and get yourself into your exam mindset. If you're well prepared, you can remind yourself of that to help keep calm. See it as a job to do and look at the test as an opportunity. If you perform well it will boost your chances of getting into good medical schools. If the worst happens, there are plenty of good medical schools that do not use UKCAT, so all is not lost.

How to Approach This Section

Time pressure is a recurring theme throughout the UKCAT, but it is especially important in Section 1, where you have only 30 seconds per question and a lot of information to take in.

Top tip! Though it might initially sound counter-intuitive, it is often best to read the question *before* reading the passage. When read the passage knowing what you're looking for, you're likely to find the information you need much more quickly.

You should look carefully to see what the question is asking. Sometimes the question will simply need you to find a phrase in the text. In other instances, your critical thinking skills will be needed and you'll have to carefully analyse the information presented to you.

Extreme Words

Words like "extremely", "always" and "never" can give you useful clues for your answer. Statements which make particularly bold claims are less likely to be true, but remember you need a direct contradiction to be able to conclude that they are false.

To answer an "always" question, you're looking for a definition. Always be a bit suspicious of "never" – make sure you're certain before saying true, as most things are possible.

Prioritise

With UKCAT, you can leave and come back to any question. **By flagging for review**, you make this easier. Since time is tight, you don't want to waste time on long passages when you could be scoring easier marks. Score the easy marks first, then come back to the harder ones if time allows. If time runs too short, at least take enough time to guess the answers as there's a good chance you could pick up some marks anyway.

Be a Lawyer

Put on your most critical and analytical hat for section A! Carefully analyse the statements like you're in a court room. Then look for the evidence! **Examine the passage closely, looking for evidence** that either supports or contradicts the statement. Remember **you're making decisions based on ONLY the passage**, not using any prior knowledge. Does the passage agree or disagree? If there isn't enough evidence to decide, don't be afraid to say "cannot tell".

Read the Question First

Follow our top tip and read the question before the passage. There is simply not enough time to read all the passages thoroughly and still have time to complete everything in 22 minutes. By reading the statement or question first, you can understand what it is that is required of you and can then pick out the appropriate area in the passage. Do not fall into the trap of trying to read all of the passage, you will not score highly enough if you do this.

When skim reading through the passage, it is inevitable that you will lose accuracy. However you can reduce this effect by doing plenty of practice so your ability to glean what you need improves. A good tip is to practice reading short sections of complicated texts, such as quality newspapers or novels, at high pace. Then test yourself to see how much you can recall from the passage.

Find the Keywords

The keyword is the most important word to help you relate the question to the passage; sometimes there might be two keywords in a question. When you read the passage, focus in on the keywords straight away. This gives you something to look for in the passage to identify the right place to work from.

It is usually easy to find the keyword/s, and you'll become even better with practice. When you find it, go back a line and read from the line before through the keyword to the end of the line after. Usually, this contains enough relevant information to give you the answer.

If this is not successful, you need to consider your next steps. Time is very tight in the UKCAT and especially so in section 1. There are other passages that need your attention, and there may be much easier marks waiting for you. If reading around the keyword has not given you the right answer it may well be time to move on. It might be that there is a more subtle reference somewhere else, that you need to read the whole passage to reach the answer or indeed that the answer cannot be deduced from the passage. Either way, if it's difficult to find your time could be better spent gaining marks elsewhere. Make a sensible guess and move on.

Use only the Passage

Your answer *must* only be based on the information available in the passage. Do not try and guess the answer based on your general knowledge as this can be a trap. For example, if the question asks who the first person was to walk on the moon, then states "the three crew members of the first lunar mission were Edwin Aldrin, Neil Armstrong and Michael Collins". The correct answer is "cannot tell" – even though you know it was Neil Armstrong and see his name, the passage itself does not tell you who left the landing craft first. Likewise if there is a quotation or an extract from a book which is factually inaccurate, you should answer based on the information available to you rather than what you know to be true.

If you have not been able to select the correct answer, eliminate as many of the statements as possible and guess – you have a 25 – 33% chance of guessing correctly in this section even without eliminating any answers, and if you've read around some keywords in the text you may well have at least some idea as to what the answer is. These odds can add a few easy marks onto your score.

Flagging for Review

There is an additional option to flag a question for review. **Flagging for review has absolutely no effect on the overall score.** All it does is mark the question in an easy way for it to be revisited if you have time later in the section. Once the section is complete, you cannot return to any questions, flagged or unflagged.

Coming back to questions can be inefficient – you have to read the instructions and data each time you work on the question to know what to do, so by coming back again you double the amount of time spent on doing this, leaving less time for actually answering questions. We feel the best strategy is to work steadily through the questions at a consistent and even pace.

That said, flagging for review has one great utility in Section 1. If you come across a particularly long or technical passage, you may want to flag for review immediately and skip on to the next passage. By coming back to the passage at the end, you allow yourself the remaining time on the hardest question. This has an advantage in each of two scenarios. If you're really tight for time, at least you maximised the time you did have answering the easier questions, thereby maximising your marks. If it turns out you have extra time to spare, you can spend it on the hardest question, allowing you a better chance to get marks you otherwise would have struggled to obtain. Thus flagging for review can be useful in Section 1, but its usefulness is probably greatest when you flag questions very soon after seeing them rather than when you have already spent time trying to find the answer.

Remember to find the right balance: if you flag too many questions you will be overloaded and won't have time to focus on them all; if you flag too few, you risk under-utilising this valuable resource. You should flag only a few questions per section to allow you to properly focus on them if you have spare time.

Worked Examples

Example 1:

In 287 BC, in the city of Athens, there lived a man named Archimedes who was a royal servant to the King. One day, the King received a crown as a birthday gift and wanted to know whether it was made of pure gold. He ordered Archimedes to find out whether the crown was indeed pure gold or an alloy. For many days, Archimedes pondered over the solution to this problem. He knew the density of gold, but could not calculate the volume of the crown.

One day, as he was bathing, he realised as he got into the bath that the volume of water displaced must be exactly equal to the volume of his own body. Upon this realisation he ran across the streets naked, yelling eureka! He weighed the crown and found its volume by immersing it in water and then calculated its density. He discovered that the density did not match that of pure gold. The crown was impure, and the blacksmith responsible for its manufacture suffered the consequences.

1. Archimedes knew the volume of the crown but could not calculate its weight
    a. True
    b. False
    c. Cannot tell

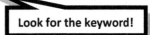

Look for the keyword!

THEN search the passage

2. Archimedes gave the crown as a birthday gift to the King
    a. True
    b. False
    c. Cannot tell
3. The crown had silver impurities
    a. True
    b. False
    c. Cannot tell
4. Archimedes found the weight of the crown using a balance scale
    a. True
    b. False
    c. Cannot tell

Answers

1. **False** – The keywords are volume and weight. Check these against the text and you will find that Archimedes could calculate the weight, but not the volume.
2. **False** – Whilst it does not explicitly state the giver of the gift, the description of Archimedes as a servant and his role in investigating the crown is wholly incompatible with him being the giver of the gift.
3. **Cannot tell** – The word silver does not appear anywhere in the passage so this statement cannot be true. But this statement is not false either because nowhere does it say that silver was not the impurity.
4. **Cannot tell** – Through your own logic, you probably guessed that this is how Archimedes weighed the crown, but remember to only use the information within the passage and use of a balance scale is not mentioned.

Example 2:

Gregor Mendel was an Austrian-Hungarian monk who is regarded to be the father of genetics. Mendel was born in poverty and was often believed to suffer from autism. He studied mathematics and physics at university, but subsequently dropped out as he could not fund his studies. He joined a monastery to escape a life of poverty. He loved to collect biological specimens and from this he noticed the different traits that animals and plants possessed. Curiosity led him to experiment with plants in a greenhouse at the monastery, as experiments using animals was forbidden.

He collected pure bred pea plants of different colours (green and yellow) and bred them together. He collected the seeds and planted them, noting that all the plants of this second generation produced green peas. He cross bred this second generation of peas and replanted the seeds. Surprisingly, of the third generation, most of the peas produced were green, and a few plants produced yellow peas. From this he deduced that the pea colour was determined by a gene that had different forms, called alleles. Using mathematics he found that the ratio of green peas to yellow peas came to 9:3:3:1, now called the classical Mendelian ratio. This work led to the development of the theory of genetics and how some alleles were dominant over the other and thus deduces the phenotype of the organism in question.

1. Gregor Mendel was a scientist
    a. True
    b. False
    c. Cannot tell

> Don't waste time! Remember to read the questions before the passage!

2. The facilities of the monastery enabled him to carry out his experiments
    a. True
    b. False
    c. Cannot tell
3. The genotype of the organism is influenced by alleles
    a. True
    b. False
    c. Cannot tell
4. The monastery allowed Mendel to carry out experiments on animals
    a. True
    b. False
    c. Cannot tell

Answers

1. **True** – This is a slightly tricky question. Mendel studied mathematics and physics, and used these skills and scientific method to produce groundbreaking scientific results. Therefore, though not explicitly stated, he is by any reasonable definition a scientist – so the statement is true.
2. **True** – It says clearly that the monastery had a greenhouse that he utilised to perform experiments on plants.
3. **Cannot tell** – Although you know this to be true from your own knowledge, it cannot be inferred from the passage.
4. **False** – This is a direct contradiction to the passage which states that experimenting on animals was forbidden at the monastery.

Example 3:

Before the 20th century, relatively little was known about the atom. The concept that objects were made of smaller particles that could not become any smaller was theorised by two Greek philosophers; Leucippus and Democritus. They believed that if you keep cutting an object consistently, there will come a point where it will not be able to be cut any further. Therefore, the theory of the atom was established but it was not possible to explore it further.

In 1897, JJ Thompson discovered the electron. He subjected a hot metal coil into an electric field, thereby producing the first cathode ray. Importantly, he noticed that the cathode ray could be deflected by a magnetic field, when viewed under a cloud chamber, and realised that it was negatively charged. As the atom is neutral, he proposed that there must be positively charged particles that give the atom an overall neutrality. JJ Thompson put forward the plum pudding model theory of the atom; that positively charged particles and negatively charged particles are mixed together in an infinitely small region of space.

In 1911, Ernest Rutherford carried out the gold leaf experiment. He fired alpha particles at a gold leaf and found that although most of the alpha particles went through, some were deflected. Occasionally, he also saw a small spark upon collision. From this, he theorised that the atom cannot be a mixture of negatively and positively charged particles, but rather has a dense core of positively charged particles. He called these particles protons. He also realised that most of the atom is empty space.

In 1932, James Chadwick performed an experiment that discovered the final component of the atom. On observation of alpha decay, he noticed that one of the particles being emitted was not deflected by a magnetic field, hence being neutrally charged. He called this particle the neutron.
Thus, the Rutherford Model of the Atom was born; the protons and neutrons form the nucleus of the atom, which electrons in spinning in orbit.

> **This is a long passage. Consider flagging for review and coming back later if you have time**

1. The passage supports which of the following conclusions?
A. The experiments of the previous scientist led to the development and guidance of the other.
B. The Rutherford Atomic Model cannot be further improved.
C. Rutherford had the help of other scientists to put forward his theory.
D. The deflection of the cathode ray by magnetism was the phenomenon that led JJ Thompson to develop the Plum Pudding model.

2. Based on the passage, each of these statements is true except
A. Earnest Rutherford is from New Zealand.
B. James Chadwick named the neutron.
C. The direction of particle deflection was determined using the cloud chamber.
D. Most of the atom is empty space.

3. Using the information in the passage, it can be inferred that:
A. Previous to Leucippus and Democritus, no one had thought of the idea of the atom.
B. Rutherford is the father of nuclear physics.
C. The gold leaf experiment was key in discovering the atomic nucleus.
D. The positron also exists.

4. Which of the following statements about the work of Earnest Rutherford is true?
A. He never carried out his own experiments without assistance from others.
B. The experimental data from the gold leaf experiment led to the development of the Geiger counter.
C. He discovered that most of the atom is empty space.
D. The foundation of nuclear fission was built from the gold leaf experiment.

Answers

1. **D** – D is the only conclusion supported by the passage, none of the other statements are mentioned in the passage. You may know from your general knowledge or simply from common sense that Rutherford had the help of other scientists, but because the passage does not mention this, it is not the answer.

2. **A** – Earnest Rutherford was indeed from New Zealand, but this is not mentioned in the passage.

3. **C** – There is no way that you know that A is true. The passage does not suggest that Rutherford is considered to be the father of nuclear physics. Finally, although the positron does indeed exist, this is not mentioned in the passage!

4. **C** – A and B are also true but not supported in the passage. D is not mentioned in the passage and also not scientifically correct.

Verbal Reasoning Questions

For questions 1 – 106 decide if each of the statements is true, false or can't tell:

SET 1

The Kyoto Protocol is an international agreement written by the United Nations in order to reduce the effects of climate change. This agreement sets targets for countries in order for them to reduce their greenhouse gas emissions. These gases are believed to be responsible for causing global warming as a result of recent industrialisation.

The Protocol was written in 1997 and each country that signed the protocol agreed to reduce their emissions to their own specific target. This agreement could only become legally binding when two conditions had been fulfilled: When 55 countries agreed to be legally bound by the agreement and when 55% of emissions from industrialised countries had been accounted for.

The first condition was met in 2002 however countries such as Australia and the United States refused to be bound by the agreement so the minimum of 55% of emissions from industrialised countries was not met. It was only after Russia joined in 2004 that allowed the protocol to come into force in 2005.
Some climate scientists have argued that the target combined reduction of 5.2% emissions from industrialised nations would not be enough to avoid the worst consequences of global warming. In order to have a significant impact, we would need to aim at reducing emissions by 60% and to get larger countries such as the US to support the agreement.

1. The Kyoto Protocol is legally binding in all industrialised countries.
2. The greenhouse gas emissions from Australia and the United States represent 45% of emissions from industrialised countries.
3. Each country chose the amount by which they would reduce their own emissions.
4. The global emission of greenhouse gases has reduced since 2005.
5. The harmful effects of climate change would be avoided if all countries reduced their emissions by 60%.

SET 2

The space race was a competition between the Soviet Union and the United States to show off their technological superiority and economic power. It took place during the Cold War when there was a tense relationship between these nations. As the technology used in space exploration could also have military applications, both nations had many scientists and technicians involved.

In 1957, the USSR launched the first artificial satellite into the Earth's orbit, named Sputnik. The launch of this satellite was one of the first steps towards space exploration. The Americans were worried that the Soviets could use similar technology to launch nuclear warheads. This prompted urgency within the Americans, leading President Eisenhower to found NASA, and so began the space race.

The Soviets took another step forward in April 1961 when they sent the first person into space, a cosmonaut named Yuri Gagarin. This prompted President John F. Kennedy to make the unexpected claim that the US would beat the Soviets to land a man on the moon and that they would do so before the end of the decade. This led to the foundation of Project Apollo, a programme designed to do this.

In 1969, Neil Armstrong and Buzz Aldrin set off for the moon on the Apollo 11 space mission and became the first astronauts to walk on the moon. Neil Armstrong famously said "one small step for man, one giant leap for mankind." This lunar landing led the US to win the space race that started with Sputnik's launch in 1957.

6. The Soviet Union were mainly concerned with launching satellites into space for a military advantage over the United States.
7. Project Apollo was founded in order for the United States to defeat the Soviet Union in the Cold War.
8. Yuri Gagarin did not become the first man on the moon because the Soviet technology could not handle the conditions on the moon.
9. The United States began their attempts at space exploration when the Soviets launched Sputnik.
10. The United States were losing the space race when John F. Kennedy said they would land a man on the moon.

SET 3

A marathon is a long distance running event that is 26.2 miles long. This race was named after the famous Battle of Marathon. The first Persian invasion of Greece took place in 490 BC. The Greek soldiers did not expect to defeat the Persian army, which had greater numbers and superior cavalry. The Greek commander utilised a tactical flank to defeat the Persians forcing them to retreat back to Asia. According to legend, the fastest Greek runner, Pheidippides, was ordered to run from Marathon to Athens to announce the Greek victory over the Persians, but then collapsed and died of exhaustion. This legendary 25 mile journey from Marathon to Athens is the basis for modern marathons.

The initial organisers of the Olympic Games in 1896 wanted an event that would celebrate the glory of Ancient Greece. They therefore chose to use the same course that Pheidippides ran. In subsequent Olympic Games, the exact length of the route depended on the location but was roughly similar to the 25 mile distance. The current standardised distance of 26.2 miles has been chosen by the IAAF and used since 1921, and has been taken from the distance used at the 1908 Olympics in London. Nowadays, more than 500 marathons are organised each year.

11. Pheidippides was chosen as he was the only Greek runner determined enough to make the journey to Athens
12. Marathon distances have been standardised since the 1908 Olympics
13. The Persian commander believed he would defeat the Greeks in the Battle of Marathon.
14. The original route from Marathon to Athens is used for IAAF marathons today.
15. The Persian soldiers were trained better than the Greek soldiers.

SET 4

Many species of bird migrate northwards in the spring to take advantage of the abundance of nesting locations and insects to eat. As the availability of food resources decreases during the winter to the point where the birds cannot survive, the birds migrate south again. Some species are capable of flying all the way around the earth. The act of migration itself can be risky for birds due to the amount of energy required to sustain flight over these long distances. Many juvenile birds can die from exhaustion during their first migration. Due to this inherent risk of migration, many species of birds have acquired different adaptations to increase the efficiency of flight. Flying with other birds in certain formations can allow their flight patterns to be more energy efficient.

The Northern bald ibis migrates from Austria to Italy. The behaviour of these birds is such that they migrate together within a flock and each individual bird continuously changes its position within the flock. Each individual bird benefits by spending some time flying in the updraft produced by the leading birds and a proportional amount of time leading the formation. Although it would theoretically be possible for an individual bird to take advantage of this energy-efficient flight without leading the formation itself, no Northern bald ibis has been shown to do this.

16. As migration is risky and dangerous, it would be better for birds not to migrate.
17. All migrating birds do so in flocks to increase their efficiency
18. All the birds within a flock of Northern bald ibis benefit from flocking behaviour
19. A bird within a Northern bald ibis flock that does not lead will be forbidden from flying with the rest of the flock
20. The migration timing depends on different seasons

SET 5

The dramatic decline of the bee population in the UK has been attributed to a number of causes such as the loss of wild flowers in the countryside. Bees require these wildflowers for food and it has been estimated that 97% of the flower-rich grassland has been lost since the 1930s. Other causes include climate change and pesticides that are toxic to bees. This is particularly problematic in the UK, which has had a 50% reduction in the honey bee population between 1985 and 2005 whilst the rest of Europe has only averaged a 20% reduction during the same time period.

This loss of flowers from the British countryside has been caused by agricultural pressures. In order to increase food production, traditional farming methods have been abandoned in favour of techniques that increase productivity. These techniques, however, involve the reduction of wild flowers.

Bumblebees are required to pollinate wildflowers and commercial crops. A reduction of wildflower pollination will result in their decline, which will ultimately affect other wildlife as they can be involved in a complex food chain. The commercial crops will need to be pollinated artificially using expensive methods that will ultimately drive up the price of fruits and vegetables. The global economic value of pollination from bees has been estimated at €265 billion annually.

21. The bee population in the UK has decreased by 97% since the 1930s.
22. The UK is the country with the largest decline in bee population
23. Reverting back to traditional farming methods will decrease the overall production of food
24. Artificial pollination will be capable of replacing bees if it becomes cheap enough
25. Increasing pesticide-free fruit and vegetable growth will slow the decline in bee population

SET 6

Driving in snowy and icy conditions can be dangerous as it increases stopping distances. The stopping distance represents how far a car will travel before slowing down to a halt. It is made up of the thinking distance, which is how long it takes for a driver to react, and the braking distance, which represents the time taken for the brakes to fully stop the car.

It is advised to fit winter tyres during the winter season as these have much better grip on snow and ice. These tyres are made out of a softer material than regular tyres, allowing them to have more traction at colder temperatures. This ultimately reduces the braking distance.

As the brakes are not very effective at stopping a vehicle on icy roads, it is recommended to steer out of trouble if possible rather than applying the brakes. It is therefore important to be travelling at slower speeds to avoid the need to suddenly brake.

If the car is stuck and cannot move, the driver can stay warm by running the engine to generate heat. However, if the exhaust pipe becomes blocked by snow and the fumes cannot escape then the engine must be turned off. This is because the engine produces carbon monoxide, which is extremely toxic and odourless.

26. Driving whilst tired increases the braking distance.
27. Regular tyres are more dangerous than winter tyres in cold conditions as they are harder
28. It is always safe to run the engine for heat in cold conditions
29. It is dangerous to use winter tyres in hot summer conditions.
30. To avoid a collision on icy surfaces, it is important to gently apply the brakes.

SET 7

The Socratic method is a form of philosophical questioning named after the Greek philosopher Socrates. It takes place as a dialogue between Socrates and another individual and attempts to investigate difficult concepts such as justice and ethics. In this dialogue, Socrates' partner puts forward an opinion or a thesis. Socrates then proposes extra premises that will attempt to disprove the original thesis. If there is an opposition, this shows that the original thesis is false and that the opposite is true.

In this dialogue, Socrates often showed other philosophers and thinkers how their reasoning was wrong. Some respected him for doing so as he was aware of the fact that his knowledge was limited and that he was merely questioning everything critically. However, many people were angered by him asking questions without providing any answers to these difficult questions himself. This led to him making a number of enemies in Greece.

In 399 BC, Socrates was accused of heresy and corruption of the youth by three of his enemies. He was then trialled and found guilty by the jury and so sentenced to death by drinking the poison hemlock. One of his friends, Crito, bribed the guards to allow Socrates to escape; however he chose not to flee away. Many have referred to him as a martyr as he chose to die standing for knowledge and wisdom.

31. Socrates was the first person to use this method of questioning.
32. The Socratic method questions the wisdom of the person forming the opinions.
33. As described in the passage, the Socratic method involved corrupting the youth and heresy.
34. Socrates chose not to escape from prison because he was afraid his enemies would find him again.
35. Socrates was able to define concepts such as justice and ethics himself.

SET 8

The Anglo-Saxons were the group of people who lived in England between the 5th Century and the Norman conquest in 1066 AD. When the Germanic tribes of the Saxons, Angles and Jutes came to Britain in 449 AD, they pushed the Celtic Britons who were there before them up into Wales. The combination of the Germanic dialects of these different tribes became Anglo-Saxon or Old English.

Old English is very different from Modern English and uses more Germanic words and its grammar is closer to Old German. If English speakers were to read a passage of Old English, they would struggle to understand any more than a few words. It is thought that this Old English is most similar to the Dutch dialect spoken in Friesland, a province in the north of the Netherlands. One of the famous literary works written in Old English was the poem Beowulf. It is not known who wrote this poem and only one original manuscript of the poem still exists today. The story involves the hero Beowulf who fights and kills the giant Grendel. All the people celebrate the death of Grendel however Grendel's mother comes to the town and attempts to kill as many people for revenge. Beowulf then fights Grendel's mother and kills her as well.

36. The Jute tribe did not contribute to the Anglo-Saxon language.
37. German speakers would be able to read Beowulf in its original language.
38. Beowulf was the strongest warrior at that time.
39. A dialect of Old English is currently spoken in the Netherlands.
40. The Celts lived in England in 449 AD

SET 9

A constellation is a group of stars that are often visible forming a pattern in the sky. The constellation's visibility depends on a number of factors. The biggest factor is your position on the Earth, for example the constellation Cassiopeia is only visible in the northern hemisphere. As the Earth orbits the Sun, another significant factor affecting constellation visibility is the season on Earth.

As the Earth rotates on its own axis, certain stars and constellations can appear to rise and fall in the night sky. Constellations that do not move in this manner are called circumpolar. This factor can allow people to navigate on the Earth using the positioning of the stars. This is especially useful during marine navigation as there are no visible landmarks. The most commonly used star for navigation is the North Star Polaris, as its position is constant within the night sky.

There are 12 constellations that take the form of animals or humans known as the zodiac signs. This is the basis for the origin of star signs in astrology, which suggests that human behaviour is influenced by the celestial phenomena. The star sign of a person represents the position of the sun at the moment of their birth. The zodiac sign which shares the same position as the sun in the sky becomes their star sign.

41. Certain stars are not visible from the southern hemisphere
42. The constellation Cassiopeia is circumpolar
43. Star signs are chosen by the constellations visible at birth
44. Polaris is used for navigation as it is the brightest star in the sky
45. Different constellations are visible from London and Sydney.

SET 10

Maple syrup is a sweet syrup made from maple trees. The sap from the trees is harvested during March. It is then boiled to evaporate water, making it denser and sweeter. It takes roughly 40 litres of maple sap to produce 1 litre of maple syrup. This current process is very similar to that used by the Native Americans except it uses more advanced equipment.

According to American Indian legend, the maple trees originally made life free from hardship. They produced a thick syrup all year round which the people would drink. A mythological creature named Glooskap saw that the people of a village were strangely silent. The men were not getting ready to hunt and the women were not minding the fires. He found the villagers sitting near the maple trees letting its syrup drip into their mouth.

Glooskap was angered by their laziness and used his powers to fill the trees with water so that they would only produce a dilute, watery sap. This meant that the people had to boil the sap to produce the sweet syrup. Although it wasn't very difficult to do this, it meant that they had to look after their fires and gather firewood. Furthermore, it meant that the trees were not able to produce enough sap to sustain the people all year so they would be forced to hunt and forage during the spring and summer.

46. Sweet syrup can be made from the sap from other trees.
47. Glooskap was angered because there was no syrup left for him to drink.
48. The technique for making maple syrup is similar to that used in the time of the American Indian legends.
49. Sap is only harvested from maple trees for one month a year.
50. Hunting animals was more difficult than drinking maple syrup.

SET 11

Stockholm syndrome is an interesting phenomenon that sometimes happens to people who have been kidnapped or held hostage. They may feel some loyalty or attraction towards the person that has kidnapped them.

This phenomenon was named after a bank robbery in Stockholm, Sweden in 1973. Four bank workers were kept hostage by two criminals who wanted to rob the bank. After being held against their will for six days, they showed that they had formed a positive relationship with their captors. The hostages were seen to be hugging and kissing the men who had kidnapped them.

It can be hard to explain why this might be the case as it involves the captor putting the hostage in a terrifying situation. In the mind of the hostage it is this person who can ultimately decide if the hostage is going to die. As a result of this fear of death, any small act of kindness prompts them to be thankful for the gift of life. This can ultimately lead to them developing Stockholm syndrome.

It is important to understand the interaction between victims and captors in cases of Stockholm syndrome as this knowledge can improve the chances of hostage survival. As such, the FBI is willing to devote resources in order to improve crisis negotiation.

51. People who get kidnapped will eventually develop positive emotions towards their kidnapper.
52. The bank workers hugged and kissed their kidnappers because the criminals forced them to.
53. Stockholm syndrome is useful to understand for crisis negotiation.
54. According to the passage, it is the victims that fear for their life that tend to develop Stockholm Syndrome.
55. The bank robbery in 1973 was the first recorded instance of a positive relationship developing between captors and hostages.

SET 12

Clownfish are a type of fish that live in salt water. They are orange fish that have white stripes. They are capable of growing up to 10-18 cm. They have also been called anemone fish as they have a symbiotic relationship with certain sea anemones.

The tentacles of sea anemone are capable of stinging fish that come near them, which the anemone then eat. The clownfish have a special mucous covering that protects them from this sting. As a result of this, clownfish that live inside sea anemone are safe from other predators but do not get stung themselves. The sea anemone benefit from the clownfish as the clownfish eat the algae that grows on the anemone. The bright colour of the clownfish lures in small fish to the anemone which ultimately get stung. They also receive better water circulation from the action of the clownfish fins.

Many people like keeping clownfish in aquariums because of their bright orange colour and how easy they are to look after. This can be dangerous as the clownfish lifespan is greatly increased by living within an anemone. Even if an anemone is added to the aquarium, only certain species of clownfish are capable of living within certain species of anemone.

56. A clownfish that loses its mucous layer can live inside sea anemone.
57. Clownfish help sea anemone to eat fish.
58. Clownfish will die if they do not have an anemone to live in.
59. Clownfish have a mutually beneficial relationship with all anemone.
60. Anemone protect clownfish from predators.

SET 13

The guillotine was a machine used to kill people by chopping off their head. It consists of a heavy blade attached to a frame. When the blade was released, it would fall down under its own weight and chop off the victim's head, killing them instantly. It was commonly used in France during the French Revolution as it was the only legal method of execution in order to enact the death penalty.

The guillotine was named after a French doctor called Joseph Guillotin. He decided that a more humane way of executing someone was needed as he realised he was unable to get rid of the death penalty. He decided that using an automatic mechanical device for decapitation would be more humane than by a person with an axe. The first guillotine built was then tested on animals to see if the axe would have enough force to decapitate its victim.

Before the guillotine was used in execution, the criminal would be hanged. This was seen to be less humane as the victim was supposed to die from the impact of the rope snapping their necks, however this did not happen all the time. The victim could be in agony for up to forty minutes before eventually dying from asphyxiation.

61. Joseph Guillotin agreed with the death penalty.
62. According to the passage, the guillotine was used for execution before the French Revolution.
63. Hanging usually causes death by asphyxiation.
64. The guillotine was seen as being more humane as it was automatic with the blade falling through its own weight.
65. During the French Revolution, hanging was the common form of execution.

SET 14

"The Gherkin" is a building in the City of London named after and famous for its distinctive shape. Its modern architecture was designed by Norman Foster and was built between 2001 and 2003. Norman Foster is famous for utilising the laws of physics in designing many of his buildings. The walls of the Gherkin would allow air to enter the building for passive cooling. As this air warms up, it rises and is then let out of the building.

The site where the Gherkin is built used to belong to the Baltic Exchange, the headquarters of the global marketplace for ship sales. In 1992, the site was damaged by bombs placed by the Provisional IRA. As there were many historic buildings in the area and only a few were damaged, the City of London governing body was insistent that any redevelopment must restore the old historic look. They later discovered that the amount of damage caused was too severe and so removed this restriction.

The building was sold in 2007 for a sum of £630 million, making it the most expensive office building in the UK. The building was then put up for sale in 2014, initially at a lower price as its owners could not afford to pay loan repayments due to high interest rates and the devaluing of the British pound. It was bought by a Brazilian billionaire for £700 million.

66. According to the passage, The Gherkin is famous because of its passive cooling system.
67. The Baltic Exchange was built with a modern architectural design.
68. The Gherkin was sold in 2014 at a loss.
69. The Provisional IRA intended to destroy the Baltic Exchange.
70. Norman Foster has designed many buildings that incorporated physics in their design.

SET 15

The living wage represents the minimum hourly rate for employment which allows an individual to cover their basic costs of living. This wage is set at £9.15 per hour in London and £7.85 per hour in the rest of the UK. This difference in living wage between London and the rest of the UK is explained by the significant costs associated with living in London.

The minimum wage is currently £6.50 for employees aged over 21 and £5.13 for employees aged between 18 and 20. This is lower than the living wage values quoted above and represents the legal minimum wage. It would be illegal for an employer to pay less than these values. Although many employers have agreed to pay a living wage, they are not legally obliged to do so.

The introduction of the living wage in companies has been argued to be beneficial for these companies. Many employers found that employees who were paid the living wage were able to work harder with better quality. It has also improved the quality of life for the families of employees as it has allowed those who are parents to spend more time with their children. However, some employers have argued that the introduction of the living wage would force them to fire some workers, forcing others to work harder.

71. The living cost across UK cities is approximately equal.
72. Companies introduce the living wage to avoid the legal complications of underpaying their employees.
73. A 24 year old working at minimum wage would be able to live comfortably in Manchester
74. Introducing a living wage would be beneficial for all employees.
75. Paying someone a living wage will only benefit that individual

SET 16

The ecological footprint is a simple way to look at how sustainable people are being. It is based on the idea that all the resources taken from the Earth are finite. It is defined by how much land and water would be required to produce the resources that the population consumes within a year. It has been calculated that there are currently 11.2 billion bio-productive hectares available on the Earth.

A study in 2004 has suggested that our ecological footprint is 13.5 billion hectares, meaning that we are using the Earth's resources 20% faster than they are being renewed. This will ultimately result in the loss of all the Earth's resources.

Individual countries can look at their own ecological footprint and compare it to the size of their bio-productive capacity. Some countries are in an ecological deficit - they require more land than their bio-productive capacity to sustain them. Other countries have an ecological reserve, meaning that their bio-productive capacity is greater than their footprint.

It can be difficult for individual countries to reduce their ecological footprint. This can either be performed by reducing that countries reliance on unsustainable resources or by increasing the amount of bio-productive land available. It is important, however, that a global effort is made to increase our sustainability.

76. The ecological footprint of a country directly depends on its population.
77. The combined area of the Earth's land and water mass is 11.2 billion hectares
78. The ecological footprint can be reduced by using sustainable resources.
79. Only the countries with an ecological deficit are able to tackle the global problem of sustainability.
80. The bio-productive capacity of the Earth is fixed.

SET 17

English punk rock band The Sex Pistols formed in 1975, and sparked off the British punk movement, leading to the subsequent creation of multiple punk and alternative rock acts. They produced only one album - 'Never Mind the Bollocks, Here's the Sex Pistols' – in their brief existence. However, despite these, some state The Sex Pistols are one of the most important bands in the history of popular music.

Originally, the band was made up of Johnny Rotten (the singer), Paul Cook (the drummer) and Glen Matlock (the bassist), however the latter was subsequently replaced by Sid Vicious. The band was involved in numerous controversies, due to their lyrics, performances and public appearances. One may argue the band was asking for trouble when they created songs attacking the music industry (such 'EMI') and commenting on controversial topics like consumerism, the Berlin wall, the Holocaust and abortion ('Bodies'). The band did not take kindly to local figureheads, as demonstrated with the release of 'God Save the Queen' in 1977, which was an attack on both conforming to societal norms and also blindly accepting the royalty as an authority.

81. It is believed by some that pop music's most influential act is The Sex Pistols.
82. The Sex Pistols kick-started the punk movement.
83. The Sex Pistols' lyrics denounced abortion.
84. The Sex Pistols controversially equated the music industry with the Holocaust.
85. The Sex Pistols' songs showed they were staunch royalists.
86. The Sex Pistols had two bassists.

SET 18

Cancer is a disease of the tissues of the body which occurs when cells begin to grow and replicate rapidly. These cells are capable of becoming masses called tumours which can obstruct the parts of the body in which they are found. Some of these tumours are also able to spread to other parts of the body in a process known as metastasis.

The incidence of cancer has significantly increased recently in the Western developed world. The most common form of cancer in the UK is breast cancer, even though it very rarely affects men. Most cancers seem to have no obvious cause whilst a few are known to have a specific cause. Mesothelioma, for example, is a type of lung cancer that is caused by exposure to asbestos. People who have worked with asbestos without adequate protection in the past are eligible for compensation from the government.

Before the 19th century, the only way to treat a cancer was to physically remove the tumour from the body. This usually involved amputation or removing a lot of tissue. It was only in the 19th century when improvements in surgical hygiene enhanced the success rates of tumour removal. At this time, Marie and Pierre Curie discovered the first non-surgical treatment, which involved irradiating the tumour.

87. Everyone exposed to asbestos will eventually suffer from mesothelioma.
88. Breast cancer is the most common type of cancer in British women.
89. The surgical success rate for tumour removal in the 19th century was improved by irradiating the tumour.
90. Irradiation is more effective than surgery in removing tumours.
91. Only cells that are capable of spreading cause tumours.

SET 19

The indigenous Australians are the native people of Australia, also known as the Aborigines. It is believed that they arrived in Australia 50,000 years ago from South-east Asia. They lived a traditional life, which involved living in wooded areas and hunting animals with spears and boomerangs. This lifestyle would not harm the environment of Australia as they believe that the land, with all of its animals and plants, is sacred.

When the British people came to Australia in 1788, there were not many people living in Australia. According to British law, any land could be claimed for the monarchy if they believed that nobody owned it and so they claimed all of the land. It was later agreed by the Australian government in 1976 that the aboriginal people would have the rights to the land where they were originally located if they could prove that they have been living there.

Nowadays, many of the 517,000 aboriginal people in Australia have chosen to integrate with the modern ways of life. They live in cities and towns and some of them have acquired professional jobs. Others have decided to maintain their traditional aboriginal way of living. Few have been unfortunate in that they haven't been educated enough to benefit from Australian society but have also lost their traditional aboriginal ways.

92. The British people claimed the Australian land because the Aborigines only lived in the wooded areas.
93. There are 517,000 Aborigines currently living in the cities and towns of Australia.
94. The aboriginal people were able to reclaim all their land in 1976.
95. Most Aborigines have received enough education to integrate with Australian culture.
96. The Aborigine lifestyle is similar to that of the south-east Asians from 50,000 years ago.

SET 20

Chilli peppers are a type of fruit grown all over the world. Most of these peppers are spicy and used in cooking. These peppers are believed to initially originate from Bolivia. As they spread throughout Europe, it was found that their spiciness had a similar taste to black peppercorns, which were incredibly valuable in the 15th century. As such, many people began to grow chilli peppers, which spread rapidly to India and China.

These chilli peppers began to be incorporated into the cooking of different cultures, being used more commonly in hotter countries. There are a number of reasons suggested for why these countries would prefer spicy food. One suggestion is that spicy food would help people to sweat in hot conditions, allowing them to cool down. This however, is unlikely to explain the prevalence of chilli peppers as it is possible to sweat without eating them. Adding spice to food was also found to be a convenient method of preserving food as it discouraged the growth of bacteria.

The peppers are spicy because they produce the chemical capsaicin. This chemical binds to receptors in the mouth and throat and cause the sensation of heat. This receptor is also capable of responding to heat directly. This is an example of labelled line coding which suggests that the perception of a stimulus is determined by which receptor it activates, not the nature of the stimulus itself.

97. Chilli peppers became more valuable in the 15th century.
98. Eating chilli peppers in hot conditions causes sweating.
99. All chilli peppers are spicy.
100. An individual lacking the capsaicin receptor would be able to eat the spiciest chillies without any pain.
101. Different methods of activating the capsaicin receptor (TRPV1) cause different sensations.

SET 21

The social determinants of health are factors in which people are born, grow, live, work and age. These factors depend on the social distribution of resources and can affect the life expectancy and quality of life that people have. These factors refer to the social context that people live in and include characteristics such as level of education, culture, stress and socioeconomic conditions as well as many others. It can be difficult to see how some of these factors may affect somebody's health but it is likely to involve a combination of multiple factors.

A number of mechanisms acting together have been described to explain why someone with a poor diet would have a lower life expectancy. One such mechanism is as follows: This individual would be more likely to be malnourished which would weaken their immune system. This would ultimately increase their likelihood of suffering from infections. Recurrent infections would place strain on the body, which could eventually result in organ failure. On average, the life expectancy in the most deprived areas can be up to 10 years lower than in the least deprived areas.

Many of these health inequalities are avoidable. Understanding how these social determinants affect the health of a population will allow us to improve its care. Knowledge of how social policies impact health can better allow us to monitor changes in health care, and is allowing us to reduce the gaps between those with a more disadvantaged social background from those with a privileged background.

102. Dropping out of school can have an effect on an individual's health.
103. The effects of the social determinants of health are principally measured by life expectancy.
104. The health impact of poor diet can be avoided completely by using multivitamins that supplement the
    immune system.
105. Nothing can be done to improve the health care of the poor in comparison to the rich.
106. Alan lives in a less deprived area than Henry so Alan will live longer than Henry.

SET 22

J.S. Mill describes his ethical theory and the reception of this in his book, 'Utilitarianism', and states:

'The creed which accepts as the foundation of morals, Utility, or the Greatest Happiness Principle, holds that actions are right in proportion as they tend to promote happiness, wrong as they tend to produce the reverse of happiness. By happiness is intended pleasure, and the absence of pain; by unhappiness, pain, and the privation of pleasure. To give a clear view of the moral standard set up by the theory, much more requires to be said; in particular, what things it includes in the ideas of pain and pleasure; and to what extent this is left an open question. But these supplementary explanations do not affect the theory of life on which this theory of morality is grounded—namely, that pleasure, and freedom from pain, are the only things desirable as ends; and that all desirable things are desirable either for the pleasure inherent in themselves, or as means to the promotion of pleasure and the prevention of pain.

Now, such a theory of life provokes in many minds, and among them in some of the most estimable in feeling and purpose, dislike. To suppose that life has (as they express it) no higher end than pleasure—no better and nobler object of desire and pursuit—they designate as utterly mean and grovelling; as a doctrine worthy only of swine, to whom the followers of Epicurus were, at a very early period, contemptuously likened; and modern holders of the doctrine are occasionally made the subject of equally polite comparisons by its German, French, and English assailants.

When thus attacked, the Epicureans have always answered, that it is not they, but their accusers, who represent human nature in a degrading light; since the accusation supposes human beings to be capable of no pleasures except those of which swine are capable.'

107. How do the Epicureans answer their critics?
A. By claiming the critics are miserable, in their refusal to embrace pleasure.
B. That the critics do not understand the multiplicity of things contained in the word 'pleasure'.
C. By calling their critics degraded.
D. By suggesting their critics are more susceptible to animalistic pleasures than they.

108. Which of the following actions are NOT in keeping with the theory of utility?
A. Providing a crash mat for a gymnast, to prevent him or her hurting him or herself.
B. Getting a crash mat for yourself, to prevent hurting yourself when performing gymnastics.
C. Not eating a chocolate bar because social pressures deem it wrong.
D. Eating a chocolate bar because it is delicious.

109. Which of the following does the passage suggest about critics of utilitarianism?
A. They are Christians.
B. They are European.
C. They are unintelligent.
D. They are reactionary.

110. Utilitarianism is only concerned with ends.
A. True
B. False
C. Can't tell

111. The above passage defines:

A. What is included by the term pleasure
B. What is included by the term pain
C. What is meant by Epicureanism
D. What is meant by utility

SET 23

Geology deals with the rocks of the earth's crust. It learns from their composition and structure how the rocks were made and how they have been modified. It ascertains how they have been brought to their present places and wrought to their various topographic forms, such as hills and valleys, plains and mountains. It studies the vestiges, which the rocks preserve, of ancient organisms that once inhabited our planet. Geology is the history of the earth and its inhabitants, as read in the rocks of the earth's crust.

To obtain a general idea of the nature and method of our science before beginning its study in detail, we may visit some valley, on whose sides are rocky ledges. Here the rocks lie in horizontal layers. Although only their edges are exposed, we may infer that these layers run into the upland on either side and underlie the entire district; they are part of the foundation of solid rock found beneath the loose materials of the surface everywhere.

Take the sandstones ledge of a valley. Looking closely at the rock we see that it is composed of myriads of grains of sand cemented together. These grains have been worn and rounded. They are sorted also, those of each layer being about of a size. By some means they have been brought hither from some more ancient source. Surely these grains have had a history before they here found a resting place—a history which we are to learn to read.

The successive layers of the rock suggest that they were built one after another from the bottom upward. We may be as sure that each layer was formed before those above it as that the bottom courses of stone in a wall were laid before the courses which rest upon them.

112. Based on the passage, each of these statements can be verified, EXCEPT?
A. We can learn about earth's inhabitants through its crust.
B. Individual layers of sandstone form one after another.
C. Rocks are made of sand.
D. Geology does not always demand explicit evidence.

113. Wall-building is used in this passage to help us understand:
A. Mountains
B. Valleys
C. Hills
D. Plains

114. The sand mentioned in the passage comes from:
A. An ancient beach
B. The sea
C. The earth's crust
D. It is undisclosed

115. A foundation of rock is **NOT** found underneath:
A. Upland
B. Lowland
C. Nowhere
D. Water

116. 'Grains of sand' are described as sorted by:
A. Shape
B. Texture
C. Age
D. Measurements

SET 24

The genus of plants called Narcissus, many of the species of which are highly esteemed by the floriculturist and lover of cultivated plants, belongs to the Amaryllis family (Amaryllidaceæ.) This family includes about seventy genera and over eight hundred species that are mostly native in tropical or semi-tropical countries, though a few are found in temperate climates.

Many of the species are sought for ornamental purposes and, on account of their beauty and remarkable odour, they are more prized by many than are the species of the Lily family. In this group is classed the American Aloe (Agave Americana) valued not only for cultivation, but also by the Mexicans on account of the sweet fluid which is yielded by its central bud. This liquid, after fermentation, forms an intoxicating liquor known as pulque. By distillation, this yields a liquid, very similar to rum, called by the Mexicans mescal. The leaves furnish a strong fibre, known as vegetable silk, from which, since remote times, paper has been manufactured. The popular opinion is that this plant flowers but once in a century; hence the name 'Century Plant' is often applied to it, though under proper culture it will blossom more frequently.

117. Which of the following are **NOT** mentioned as potential uses for a narcissus plant:

A. Perfume production
B. Alcohol production
C. Visual decoration
D. Stationary production

118. Why is the plant known as 'the century plant'?

A. It is sown only once every hundred years.
B. It can only able to be fertilised once a century.
C. It is perceived as blooming centennially.
D. It can only able to flower once within a hundred years.

119. Which of the following statements is most supported by the above passage:

A. Lilies are generally valued less than members of the Narcissus genus.
B. Lilies are famously not as attractive as members of the Narcissus genus.
C. A number are people prefer members of the Narcissus genus over Lilies.
D. Members of the Narcissus genus are a welcome addition to any household.

120. Which of the following statements is NOT true:

A. American Aloe can be used to make rum.
B. The Amaryllis family contains more than six hundred species of Narcissus.
C. Members of the Narcissus genus can be found in all climates.
D. The members of the Narcissus genus have a distinctive smell.

121. Which of the following statements can be verified by the passage:

A. The 'Narcissus' genus is named after the mythical character, famed for his beauty.
B. Agave syrup can be collected by American Aloe.
C. A genus belongs to a family.
D. Members of the Narcissus genus are used for their soothing properties.

SET 25

The following passage is found in a book on nature published in 1899:

Five women out of every ten who walk the streets of Chicago and other Illinois cities, says a prominent journal, by wearing dead birds upon their hats proclaim themselves as lawbreakers. For the first time in the history of Illinois laws it has been made an offense punishable by fine and imprisonment, or both, to have in possession any dead, harmless bird except game birds, which may be possessed in their proper season. The wearing of a tern, or a gull, a woodpecker, or a jay is an offense against the law's majesty, and any policeman with a mind rigidly bent upon enforcing the law could round up, without a written warrant, a wagon load of the offenders any hour in the day, and carry them off to the lockup. What moral suasion cannot do, a crusade of this sort undoubtedly would.

Thanks to the personal influence of the Princess of Wales, the osprey plume, so long a feature of the uniforms of a number of the cavalry regiments of the British army, has been abolished. After Dec. 31, 1899, the osprey plume, by order of Field Marshal Lord Wolseley, is to be replaced by one of ostrich feathers. It was the wearing of these plumes by the officers of all the hussar and rifle regiments, as well as of the Royal Horse Artillery, which so sadly interfered with the crusade inaugurated by the Princess against the use of osprey plumes. The fact that these plumes, to be of any marketable value, have to be torn from the living bird during the nesting season induced the Queen, the Princess of Wales, and other ladies of the royal family to set their faces against the use of both the osprey plume and the aigrette as articles of fashionable wear.

122. In 1899:
A. Women across the USA could be prosecuted for owning ornamental dead birds.
B. There was a significant rise of female arrests in America.
C. Possession of a dead gull could lead to trouble.
D. Americans responded to law by citing the use of jays as ornamentation unfashionable.

123. Ostrich feathers were seen as preferable to osprey plums because:
A. Ostriches are less intelligent birds.
B. Ostriches are killed for their meat, so one might as well use their feathers.
C. Queen Elizabeth has an especial love of ospreys.
D. Harvesting osprey feathers was seen as an inhumane process.

124. Games birds could be possessed by citizens of Illinois all year round.
A. True
B. False
C. Can't tell

125. Banning Osprey feathers in the UK's army was difficult because:
A. Many uniforms required them.
B. The Princess did not have the authority to implement the ban.
C. Her ultimate support was predominately female, and thus their concerns seemed to have no relevance from the male domain of the army.
D. It would be hard to differentiate between other regiments within the army, who were already wearing ostrich feathers.

126. Which of the following could NOT be legally owned in Illinois, according to the passage:
A. A live bird intended for personal ornamentation.
B. A dead bird of prey that had violently attacked you.
C. Feathered garments.
D. None of the above.

SET 26

Indie game developer Lucas Pope created 'Papers Please', a video game where the player is an immigration officer processing people attempting to enter Arstotzka, a fictional dystopia. Released in 2013, the game was originally made for Microsoft Windows and OS X platforms. It was subsequently released for Linux and the iPad in 2014.

The game is set in 1982, and gameplay involves the player processing large numbers of applicants attempting to enter the country, through checking various pieces of paperwork. This is intended to keep criminals out, whether they are terrorists or drug smugglers. When looking through the applicant's 'papers', discrepancies may be discovered: the player must then enquire about these and may go on to use other tools, such as a body scanner and finger printing to discover the truth of the candidate's motives. Applicants may attempt to bribe the officer in order to get through. Ultimately, the game player must stamp candidates passports, either accepting into or rejecting them from the country. Their work is being monitored, however: after two false acceptances/rejections, the player will be pecuniarily punished, with their day's wages being decreased in response to their administrative sloppiness. They have a limited amount of time, representing each 'day', to work, during which they will be paid in accordance to the number of people processed.

127. Which of the following statements is best supported by the above passage:

A.   Lucas Pope created the Papers Please for a small games company.
B.   Papers Please is a multi-platform game.
C.   Arstotzka is a fictionalised version of an ex-Soviet block state.
D.   The game gained significant media attention in 2014.

128. Which of the following statements best sums up the official job of the player's character:

A.   To accept as many applicants into Artstotzka as possible.
B.   To reject as many new applicants entering Arstotzka as possible.
C.   To avoid making mistakes in processing people.
D.   To stamp passports.

129.   Discrepancies in information provided by applicants lead to the player:

A.   Interrogating and performing a fingerprint check on the suspicious individual.
B.   Interrogating and performing a full body scan on the suspicious individual.
C.   Asking the suspicious individual for further information.
D.   Performing one or multiple physical assessments of the individual.

130.   Which of the following statements is true:

A.   The game-player solely makes money through processing applicants.
B.   The game-player will ultimately be responsible for multiple arrests.
C.   The game-player will not be forgiven for their mistakes.
D.   The game-player may be subject to fiscal penalisation.

SET 27

Emerging in 1970s USA, Blaxploitation, or 'blacksploitation', gives homage to many other genres: within it, there are western, martial arts films, musicals, coming-of-age dramas and comedies, and the genre has even parodied itself with films like 'Black Dynamite'. Blaxploitation movies may take place in the South, and focus on issues like slavery, or be set in the poor neighbourhoods of the Northeast or West coast, but in any case they will feature a predominately black cast. It is also known to feature soundtracks comprised of soul and funk music, and the common feature of character's using the words 'honky', 'cracker' and other slurs against white people.

Originally, the genre's exports were aimed at city-dwelling black Americans, but their appeal has since grown and is not exclusive to any race. Despite the negative sound of the title 'blaxploitation', the term was coined by ex-film publicist Junius Griffin, the then head of LA's NAACP, National Association for the Advancement of Coloured People. He came up with the name through a play on the word 'sexploitation' describing films which featured pornographic scenes.

The film 'Shaft' and 'Sweet Sweetback's Baadasssss Song' are two of the forerunners of this genre, both released in 1971. The latter has been said, by Variety, to have created the genre

131. Which of the following statements is supported by the information in the above passage:

A.   'Blaxploitation' was a term coined by porn directors moving into a new genre.
B.   'Blaxploitation' was a term made popular by black audiences.
C.   'Blaxploitation' was a term coined by a civil rights activist.
D.   'Blaxploitation' was a term criticised by white sympathisers.

132. Which of the following statements best describes the most common element of a Blaxploitation film:
A.   Characters performing funk songs.
B.   Characters coming to terms with the legacy of slavery.
C.   Characters performing martial arts.
D.   Characters using racial slurs.

133. Which of these statements best describes casting in Blaxploitation films:

A.   Primarily white
B.   Primarily black
C.   Exclusively white
D.   Exclusively black

134. Which of the following best describes the audiences of Blaxploitation films:

A.   Originally for all middle-class African-Americans.
B.   Originally for all urban-dwellers.
C.   Multi-ethnic.
D.   Shrinking since the mid-1970s.

135. The legacy of films including soft-core porn is knowingly acknowledged in the name of two Blaxploitation titles, 'Shaft' and 'Sweet Sweetback's Baadasssss song', both of which suggest body parts associated with sex films.
A.   True
B.   False
C.   Can't tell

SET 28

When discussing his famous character Rorschach, the antihero of 'Watchmen', Moore explains, 'I originally intended Rorschach to be a warning about the possible outcome of vigilante thinking. But an awful lot of comic readers felt his remorseless, frightening, psychotic toughness was his most appealing characteristic – not quite what I was going for.' Moore misunderstands his own hero's appeal within this quotation: it is not that Rorschach is willing to break little fingers to extract information, or that he is happy to use violence, that makes him laudable. The Comedian, another 'superhero' within the alternative world of Watchmen, is a thug who has won no great fan base; his remorselessness (killing a pregnant Vietnamese woman), frightening (attempt at rape), psychotic toughness (one only has to look at the panels of him shooting out into a crowd to witness this) is repulsive, not winning. This is because The Comedian has no purpose: he is a nihilist, and as a nihilist, denies any potential meaning to his fellow man, and so to the comic's reader. Everything to him is a 'joke', including his self, and consequently his own death could be seen as just another gag.

Rorschach, on the other hand, does believe in something: he questions if his fight for justice 'is futile?' then instantly corrects himself, stating 'there is good and evil, and evil must be punished. Even in the face of Armageddon I shall not compromise in this.' Jacob Held, in his essay comparing Rorschach's motivation with Kantian ethics, put forward the postulation 'perhaps our dignity is found in acting as if the world were just, even when it is clearly not.' Rorschach then causes pain in others not because he is a sadist, but because he feels the need to punish wrong and to uphold the good, and though he cannot make the world just, he can act according to his sense of justice - through the use of violence.

136. Which of the following best describes 'Watchmen':
A. A book that contains only vicious characters.
B. An expression of despair when contemplating an imperfect world.
C. An example of how an author's intentions are not always realised.
D. A book that accidentally glamorises violence.

137. 'The Comedian' is a misnomer - the character that goes by this title should not, logically, be called this.
A. True
B. False
C. Can't tell

138. Which of the following best articulates the view put forward by Jacob Held?
A. We find dignity through just actions.
B. If one decides to behave as though the world is fair, this may lead to a discovery of self-worth.
C. It is shameful to view the world as corrupt.
D. Self-value can only be found in madness.

139. What does the passage above argue?
A. Rorschach breaking little fingers is preferable to the Comedian attempting rape somebody.
B. The Comedian's depressing sense of humour has made him unpopular.
C. Rorschach is not actually violent.
D. Rorschach is popular because his aggressive behaviour has a moral intent, and is not just violence.

140. What does the word 'nihilist' mean in the context of the passage?
A. Someone who believes there is no meaning to life.
B. Someone who is full of anger at the corruption of society.
C. Someone who is narcissistic.
D. Someone who hates other people.

SET 29

'The Bechdel Test', also known as the 'Mo Movie Measure' and 'The Bechdel Rule' is named after cartoonist Alison Bechdel, who in 1985 wrote a cartoon containing the original proposal of the 'test'. It depicts one woman telling another that she has 'a rule' that she will only see a film if it satisfies three basic requirements: that it contains at least two women, that they talk to each other and that their conversation is on something other than a man. The second woman states that this is 'pretty strict, but a good idea', to which the first responds the last film she saw that complied with this was 'Alien'. The original notion described in the strip has been attributed to Liz Wallace, and the test is sometimes referred to as the Bechdel/Wallace Test.

Following this, a website entitled 'The Bechdel Test Movie List' has formed an extensive list of cinematic output, showing movies that pass and do not pass the test. One may be surprised at the number of movies that would not be watched by first woman in the Bechdel comic: many blockbusters do not make the mark, and such titles as 'Godzilla', 'The Imitation Game' and 'Robocop' all feature on the list of 'failed' films.

141. According to The Bechdel Test, 'The Imitation Game' is a sexist film.

A. True
B. False
C. Can't tell

142. Which of the following films would pass the Bechdel test:

A. One where the only conversation between two women is on woman A's brother.
B. One where there are two conversations, one on woman A's son and another on woman B's boss, Mr Smith.
C. One where there are five women, in which at one point they all have a chat about how to lose weight and the best hair removal techniques.
D. One where there is one woman who chats about all manner of things with her male colleagues, including her USA presidential campaign.

143. 50% of horror films, according to the above extract, pass the Bechdel test.

A. True
B. False
C. Can't tell

144. Which of the following phrases best describes the reaction of the second woman within the comic strip:

A. Ecstatically approving
B. Condemning
C. Cautiously approving
D. Apathetic

145. The two women in the comic strip are manifestations of Bechdel and Wallace, with the piece of art being a recreation of their original conversation on this matter.

A. True
B. False
C. Can't tell

SET 30

There are many comic tropes a comedian or group of comedians may want to employ in their set or act, but for the purpose of this extract we shall focus on the device of the 'call-back'. A call-back is a reference made to a previous joke, in a different context: for example, a comedian may make the joke 'why did the chicken cross the road? To get to the other side' early on in his or her set, and then later on may reference this again by telling an anecdote and saying 'so then I crossed the road - oh, look, there's a chicken! Strange, I could have sworn he was over there a moment ago'. Though the call-back may appear to simply rely on the idea that repetition is inherently funny, it actually has several desirable effects. Firstly, it means that one joke can provide more than one laugh, as the memory of the previous joke encourages renewed chuckling, and so the original quip's comic potential is increased. It also builds up a relationship between comedian and audience, as it builds up a sense of familiarity with the speaker and his or her subject matter, and this bond also may encourage more laughter - the second joke creates the same feeling as an 'in-joke'. If used at the end of a set - as a call-back often is - it gives a sense of completion, and also may lead to the ending of the act culminating in the largest laugh.
In TV, a call-back often refers to a joke made in a previous episode.

146. Repetition is inherently funny.

A.   True
B.   False
C.   Can't tell

147. Which of the following best explains how a call-back works:

A.   Previous understanding of a subject makes it potentially more comic.
B.   Doubling a joke makes it potentially twice as funny.
C.   Making the audience feel comfortable is more likely to make them laugh.
D.   We find people we have a relationship with funny.

148. For a call-back to work, the original joke has to be significantly funny.

A.   True
B.   False
C.   Can't tell

149. A call-back cannot be used in an un-comic setting.

A.   True
B.   False
C.   Can't tell

150. Which of the following statements is best supported by the above passage:

A.   A call-back is used to create a sense of the circle having fully come to pass.
B.   A call-back can be a useful addition to an individual comedian's set.
C.   A call-back creates unity through disparate TV episodes.
D.   A call-back is an especially important trope to consider.

SET 31

Harriet Beecher (Stowe) was born June 14, 1811, in the characteristic New England town of Litchfield, Connecticut. Her father was the Rev. Dr. Lyman Beecher, a distinguished Calvinistic divine, her mother Roxanna Foote, his first wife. Harriet Beecher was ushered into a household of happy, healthy children, and found five brothers and sisters awaiting her. The eldest was Catherine, born September 6, 1800. Following her were two sturdy boys, William and Edward; then came Mary, then George, and at last Harriet. Another little Harriet was actually born three years before, but died when aged only one month old; the fourth daughter, the subject of this passage, was named in memory of this sister Harriet Elizabeth. Just two years after Harriet was born, in the same month, another brother, Henry Ward, was welcomed to the family circle, and after him came Charles, the last of Roxanna Beecher's children.

The first memorable incident of Harriet's life was the death of her mother, which occurred when she was four years old, and which ever afterwards remained with her as the most tender, sad and sacred memory of her childhood. Mrs Stowe's recollections of her mother are found in a letter to her brother Charles, afterwards published in the 'Autobiography and Correspondence of Lyman Beecher.' She says: —

"I was between three and four years of age when our mother died, and my personal recollections of her are therefore but few. But the deep interest and veneration that she inspired in all who knew her were such that during all my childhood I was constantly hearing her spoken of, and from one friend or another some incident or anecdote of her life was constantly being impressed upon me.

151. Harriet, the main character in the article, was the third daughter of Roxanna Beecher:
A.   True
B.   False
C.   Can't tell

152. Which of the following statements, according to the passage, are true:
A.   Harriet Beecher had a religious father.
B.   Harriet Beecher was born in the English town Litchfield.
C.   Harriet Beecher was born in the 18th century.
D.   Harriet Beecher was born in an average American town.

153. Roxanna Beecher was an admired woman.
A.   True
B.   False
C.   Can't tell

154. Harriet Beecher Stowe's mother's death is described as:
A.   Her saddest memory of her life.
B.   The earliest significant event in her life.
C.   Her most tender memory of her life.
D.   All of the above.

155. Which of the following statements is supported by the above passage:
A.   Harriet Beecher was between three and four when her mother died.
B.   Harriet Beecher had five brothers waiting for her when she was born.
C.   Harriet Beecher was a letter-writer.
D.   Harriet Beecher was an autobiographer.

SET 32

Gutenberg's father was a man of good family. Very likely the boy was taught to read. But the books from which he learned were not like ours; they were written by hand. A better name for them than books is 'manuscripts,' which means handwritings.

While Gutenberg was growing up a new way of making books came into use, which was a great deal better than copying by hand. It was what is called block printing. The printer first cut a block of hard wood the size of the page that he was going to print. Then he cut out every word of the written page upon the smooth face of his block. This had to be very carefully done. When it was finished the printer had to cut away the wood from the sides of every letter. This left the letters raised, as the letters are in books now printed for the blind.
The block was now ready to be used. The letters were inked, paper was laid upon them and pressed down. With blocks the printer could make copies of a book a great deal faster than a man could write them by hand. But the making of the blocks took a long time, and each block would print only one page.

Gutenberg enjoyed reading the manuscripts and block books that his parents and their wealthy friends had; and he often said it was a pity that only rich people could own books. Finally he determined to contrive some easy and quick way of printing.

156. Which of the following reasons can be inferred from the above passage to explain Gutenberg's desire to create a new way of printing was:
A. It was a lucrative business to go into.
B. He wanted to make text more accessible.
C. He was tired of waiting for each book to be hand written or block pressed, and wanted quicker access to literature.
D. He found the current books too costly for him to continue his reading habit.

157. Which of the following of the following is **NOT** mentioned as a concern of block printing?
A. It exhausts the carver.
B. It is intricate and demands attention to detail.
C. It is a lengthy process.
D. An individual block has limited utility.

158. Which of the following statements is definitely true according to the above passage?
A. Gutenberg was taught to read as a boy.
B. Gutenberg's father belonged to the aristocracy.
C. Block printing was the predominant book manufacturing process whilst Gutenberg was growing up.
D. Gutenberg's family was somewhat sociable.

159. Printing with the block process was a simple task of inking up the prepared block and pressing it down on a piece of paper, to make one page of the text.
A. True
B. False
C. Can't tell

160. Which of the following statements are **NOT** supported by the above passage:
A. Manuscripts were beautifully crafted.
B. 'Manuscripts' is an appropriate name for what it describes.
C. Block printing is an appropriate name for what it describes.
D. Having well off friends was a good way to expand your reading.

SET 33

Cassandra may be considered an odd name to give your daughter, when you consider the mythical significance of it. Cassandra was a figure in ancient Greek mythology, a Trojan girl born to King Priam, who had been cursed: she had the gift of prophecy, but no one would believe in her words. She ultimately ends up taken from her homeland, as the sexual slave of Agamemnon. Agamemnon's wife then slaughters the girl, and one might wonder why any parent would name their daughter after such an ill-fated figure.

There are several stories that explain how Cassandra gained her gift and her curse. One narrative states that the god Apollo gave the girl the ability to tell the future, in an attempt to seduce her. When she refused him, he corrupted her gift. Another version tells us that Cassandra originally told Apollo she would have sex with him, in exchange for the gift of prophecy. When she subsequently refused him, having attained this power, he then spat in her mouth during a kiss, and this action made her ever after doomed to be disbelieved.

The figure of Cassandra has been presented in various pieces of classical literature, including Homer's epic poem 'The Iliad', Euripides' 'Trojan Women' and Aeschylus' 'The Agamemnon'. Although the presentation of her character alters in the different manifestation, the tragic fate of the woman is known within the different texts, as it would be known by the different authors and audiences of these works.

161. Throughout mythology, Apollo is always presented as the figure who gives Cassandra prophecy.

A. True
B. False
C. Can't tell

162. Which of the following statements is best supported by the passage:

A. Parents who call their daughter Cassandra must hate their children.
B. Cassandra prizes chastity higher than her personal comfort.
C. Cassandra had supernatural powers.
D. Cassandra is often seen as a home wrecker.

163. Which of the following statements is **NOT** supported by the passage:

A. Cassandra comes from a royal line.
B. Cassandra had a happy childhood before her horrible fate.
C. Cassandra has been written about for the stage.
D. Homer has been inspired by Cassandra.

164. Cassandra's personality is consistently presented in the different pieces of literature she is included in.

A. True
B. False
C. Can't tell

165. Which of the following is offered in the above passage to explain Apollo's ire:

A. Cassandra breaking her promise.
B. Cassandra not accepting his gift.
C. Cassandra demanding more gifts.
D. Cassandra not acknowledging his gifts.

SET 34

Despite the fact that some associate musicals with cheesy joy, the genre is not limited to gleeful stories, as can be demonstrated by the macabre musical, 'Sweeney Todd'. The original story of the murderous barber appears in a Victorian penny dreadful, 'The String of Pearls: A Romance'. The penny dreadful material was adapted for the 19th century stage, and in the 20th century was adapted into two separate melodramas, before the story was taken up by Stephen Sondheim and Hugh Wheeler. The pair turned it into a new musical, which has since been performed across the globe and been adapted into a film starring Johnny Depp.

Sondheim and Wheeler's drama tells a disturbing narrative: the protagonist, falsely accused of a crime by a crooked judge, escapes from Australia to be told that his wife was raped by that same man of the court. In response, she has committed suicide, and her daughter - Todd's daughter - has been made the ward of the judge. The eponymous figure ultimately goes on a killing spree, vowing vengeance for the people who have wronged him but also declaring 'we all deserve to die', and acting on this belief by killing many of his clients, men who come to his barbershop. His new partner in crime, Mrs Lovett, comes up with the idea of turning the bodies of his victims into the filling of pies, as a way of sourcing affordable meat - after all, she claims, 'times is hard'.

Cannibalism, vengeance, murder and corruption - these are all themes that demonstrate that this show does not conform to a happy-clappy preconception of its genre.

166. Which of the following statements are best supported by the above passage:
A.   Sondheim is a brilliant musician and lyricist.
B.   Most musicals deal with morbid themes.
C.   Wheeler is an avid penny dreadful fan.
D.   Generalisations can be misleading.

167. All the adjectives below are explicitly supported by the passage as ways of describing the crimes described within it, except:
A.   Comic
B.   Culinary
C.   Vengeful
D.   Sexual

168. Mrs Lovett and Sweeney Todd are in a romantic relationship.
A.   True
B.   False
C.   Can't tell

169. The best way to describe the belief of Todd as mentioned in the above passage:
A.   Bad people should die so good can live and prosper.
B.   Good people should die because the bad have basically taken over.
C.   All men should die.
D.   All humans merit death.

170. Which of the following statements is best supported in the above passage:
A.   There are four themes in 'Sweeney Todd'.
B.   Legal corruption is the predominate theme of 'Sweeney Todd'.
C.   Several 'Sweeney Todd' themes are morbid.
D.   There is nothing positive in 'Sweeney Todd'.

SET 35

The United States released the following as part of a pamphlet titled 'If Your Baby Must Travel in Wartime', released during the Second World War:

'Have you been on a train lately? The railroads have a hard job to do these days, but one that they are doing well. But before you decide on a trip with a baby, you should realise what a wartime train is like. So let's look into one.
This train is crowded. At every stop more people get on—more and still more. Soldiers and sailors on furloughs, men on business trips, women — young and not so young — and babies, lots of them, mostly small.
The seats are full. People stand and jostle one another in the aisle. Mothers sit crowded into single seats with toddlers or with babies in their laps. Three sailors occupy space meant for two. A soldier sits on his tipped-up suitcase. A marine leans against the back of the seat. Some people stand in line for 2 hours waiting to get into the diner, some munch sandwiches obtained from the porter or taken out of a paper bag, and some go hungry. And those who get to the diner have had to push their way through five or six moving cars.

You will want to think twice before taking your baby into such a crowded, uncomfortable place as a train. And having thought twice, you'd better decide to stay home unless your trip is absolutely necessary.

But suppose you and your baby must travel. Well then, you will have to plan for the dozens of small but essential things incidental to travelling with a baby and equip yourself to handle them.'

171. First World War passenger trains were exceptionally crowded.
A. True
B. False
C. Can't tell

172. Which of the following phrases is described by the above passage:
A. A soldier responds to the situation by creating his own seat.
B. A sailor rest against a seat's back.
C. Many people queue for over an hour to get to the diner car.
D. Many go without eating for the duration of a train journey.

173. The pamphlet wishes to increase the number of passengers on trains.
    True
A. False
B. Can't tell

174. Every station the described train passes through has passengers wanting to get onto the vehicle.
A. True
B. False
C. Can't tell

175. Which of the following does the above passage do:
A. Compliment the railroads.
B. Insult passengers who are mothers.
C. Insult passengers who work for the navy.
D. Compliment soldiers.

SET 36

The following extract is from 'Foods That Will Win the War', published in the USA during the First World War:

'A slice of bread seems an unimportant thing. Yet one good-sized slice of bread weighs an ounce. It contains almost three-fourths of an ounce of flour. If every one of the country's 20,000,000 homes wastes on the average only one such slice of bread a day, the country is throwing away daily over 14,000,000 ounces of flour—over 875,000 pounds, or enough flour for over a million one-pound loaves a day. For a full year at this rate there would be a waste of over 319,000,000 pounds of flour—1,500,000 barrels—enough flour to make 365,000,000 loaves.

As it takes four and one-half bushels of wheat to make a barrel of ordinary flour, this waste would represent the flour from over 7,000,000 bushels of wheat. Fourteen and nine-tenths bushels of wheat on the average are raised per acre. It would take the product of some 470,000 acres just to provide a single slice of bread to be wasted daily in every home.

But someone says, 'a full slice of bread is not wasted in every home.' Very well, make it a daily slice for every four or every ten or every thirty homes—make it a weekly or monthly slice in every home—or make the wasted slice thinner. The waste of flour involved is still appalling. These are figures compiled by government experts, and they should give pause to every housekeeper who permits a slice of bread to be wasted in her home.'

176. According to the above passage, a slice of bread:

A.   Contains 1/6 lb. of flour
B.   Contains a 1/4-ounce of air
C.   Is 75% flour
D.   Is one fourth salt, butter and yeast

177. The passage denies that 20,000,000 homes at the point of writing wasted at least a slice of bread a day.

A.   True
B.   False
C.   Can't tell

178. If 20,000,000 homes wasted a slice of bread, this waste would be equal to:

A.   One million loaves of bread a day.
B.   Over 319,000,000 bushels of flour in 365 days.
C.   1.5 million barrels per annum.
D.   Over 365,000 loaves a year.

179. According to the above passage, a slice of bread is an unimportant thing.

A.   True
B.   False
C.   Can't tell

180. Which of the following statements are supported by the above passage:

A.   The writer has received much criticism for his views.
B.   The government should do more to inform the public about waste.
C.   The government has taken responsibility for public waste.
D.   Responsibility lies with the person who keeps the house.

SET 37

At the election of President and Vice President of the United States, and members of Congress, in November, 1872, Susan B. Anthony, and several other women, offered their votes to the inspectors of election, claiming the right to vote, as among the privileges and immunities secured to them as citizens by the fourteenth amendment to the Constitution of the United States. The inspectors, Jones, Hall, and Marsh, by a majority, decided in favour of receiving the offered votes, against the dissent of Hall, and they were received and deposited in the ballot box. For this act, the women, fourteen in number, were arrested and held to bail, and indictments were found against them, under the 19th Section of the Act of Congress of May 30th, 1870, (16 St. at L. 144.) independently charging them with the offense of knowingly voting without having a lawful right to vote. The three inspectors were also arrested, but only two of them were held to bail, Hall having been discharged by the Commissioner on whose warrant they were arrested. All three, however were jointly indicted under the same statute—for having knowingly and wilfully received the votes of persons not entitled to vote.

Of the women voters, the case of Miss Anthony alone was brought to trial, a nolle prosequi having been entered upon the other indictments. Upon the trial of Miss Anthony before the U.S. Circuit Court for the Northern District of New York, at Canandaigua, in June, 1873, it was proved that before offering her vote she was advised by her counsel that she had a right to vote; and that she entertained no doubt, at the time of voting, that she was entitled to vote.

181. According to the above passage, how many people in total were arrested due to the group of women voting?

A.   Fourteen
B.   Three
C.   Seventeen
D.   Sixteen

182. Susan B. Anthony was the only person brought to trial because of the incident.

A.   True
B.   False
C.   Can't tell

183. Which of the following best describes initial opinions of the election officers:

A.   United by each member's personal support of the women's votes.
B.   Divided in response to the women's actions.
C.   Apathetic about the women's actions.
D.   United by general disapproval of the women's actions.

184. Which defence for Susan B. Anthony is mentioned above?

A.   She did not realise she was not allowed to vote.
B.   That all people born in the USA should be able to vote for their president.
C.   That gender should not prevent her vote.
D.   The election officers accepted her vote, showing the responsibility is not with her.

185. The women were charged jointly under the same indictment.
A.   True
B.   False
C.   Can't tell

SET 38

The following is taken from a book about Norway published in 1909:

'In a country like Norway, with its vast forests and waste moorlands, it is only natural to find a considerable variety of animals and birds. Some of these are peculiar to Scandinavia. Some, though only occasionally found in the British Isles, are not rare in Norway; whilst others (more especially among the birds) are equally common in both countries.

There was a time when the people of England lived in a state of fear and dread of the ravages of wolves and bears, and the Norwegians of the country districts even now have to guard their flocks and herds from these destroyers. Except in the forest tracts of the Far North, however, bears are not numerous, but in some parts, even in the South, they are sufficiently so to be a nuisance, and are ruthlessly hunted down by the farmers. As far as wolves are concerned civilization is, fortunately, driving them farther afield each year, and only in the most out-of-the-way parts are they ever encountered nowadays. Stories of packs of hungry wolves following in the wake of a sleigh are still told to the children in Norway, but they relate to bygone times—half a century or more ago, and such wild excitements no longer enter into the Norsemen's lives.'

186. Which of the following is best supported by the above passage:

A.   The variety of birds and animals to be found in Norway is unique to that country.
B.   The variety of birds and animals to be found in Norway is common to all European countries.
C.   By having forests, a country is more likely to have a variety of birds and animals.
D.   England and Norway have similar geographical features.

187. English people are described as:

A.   Having been anxious of certain animals.
B.   Sceptical of bears.
C.   Living in fear of wolves.
D.   Developmentally behind the Norwegians.

188. Bears are described as:

A.   Hunting
B.   Scavenging
C.   Damaging
D.   Man-eating

189. Bears are also:

A.   Numerous in all forest tracts.
B.   Numerous throughout the North.
C.   Numerous throughout the South.
D.   At risk in parts of Norway.

190. The passage suggests:

A.   The movement of wolves to the out-of-reach parts of Norway is beneficial.
B.   Wildlife currently threats Norwegian children.
C.   Regret at the loss of adventures.
D.   Norsemen particularly respect their natural surroundings.

SET 39

The following extract is taken from Freud's book 'Dream Psychology: Psychoanalysis for Beginners'

In what we may term pre-scientific days, people were in no uncertainty about the interpretation of dreams. When they were recalled after awakening they were regarded as either the friendly or hostile manifestation of some higher powers, demoniacal and divine. With the rise of scientific thought the whole of this expressive mythology was transferred to psychology; today there is but a small minority among educated persons who doubt that the dream is the dreamer's own psychical act.

But since the downfall of the mythological hypothesis an interpretation of the dream has been wanting. The conditions of its origin; its relationship to our psychical life when we are awake; its independence of disturbances which, during the state of sleep, seem to compel notice; its many peculiarities repugnant to our waking thought; the incongruence between its images and the feelings they engender; then the dream's evanescence, the way in which, on awakening, our thoughts thrust it aside as something bizarre, and our reminiscences mutilating or rejecting it—all these and many other problems have for many hundred years demanded answers which up till now could never have been satisfactory. Before all there is the question as to the meaning of the dream, a question that is in itself double-sided. There is, firstly, the psychical significance of the dream, its position with regard to the psychical processes, as to a possible biological function; secondly, has the dream a meaning—can sense be made of each single dream as of other mental syntheses?

191. Dreams used to be regarded as having a potentially religious quality.
A. True
B. False
C. Can't tell

192. According to the passage, at this point of time, amongst the educated:
A. A vocal majority believe that dreams come from somewhere outside the dreamer.
B. A small minority believes that dreams come from the dreamer alone.
C. The majority accepts that a dreamer's dream is his or her own psychical act.
D. A vocal minority believes dreams are the direct products of angels and devils.

193. With a dream:
A. Images seemingly logically dictate feelings.
B. Events happen which are pleasant to waking thought.
C. Only boring things occur that are often too dull to be remembered.
D. There are relationships between images and feelings that would appear illogical to the awake mind.

194. The passage wonders about the significance of individual dreams.
A. True
B. False
C. Can't tell

195. Which of the following statements is supported by the above passage:
A. There is a definite link between the waking and dreaming self.
B. Human society has never had a hypothesis to explain dreams that has satisfied them.
C. A memory of a dream may be untrustworthy.
D. The origin of the dream has been scientifically sourced.

SET 40

Most of the colonists who lived along the American seaboard in 1750 were the descendants of immigrants who had come in fully a century before; after the first settlements there had been much less fresh immigration than many latter-day writers have assumed. According to Prescott F. Hall, "the population of New England ... at the date of the Revolutionary War ... was produced out of an immigration of about 20,000 persons who arrived before 1640," and we have Franklin's authority for the statement that the total population of the colonies in 1751, then about 1,000,000, had been produced from an original immigration of less than 80,000. Even at that early day, indeed, the colonists had begun to feel that they were distinctly separated, in culture and customs, from the mother-country and there were signs of the rise of a new native aristocracy, entirely distinct from the older aristocracy of the royal governors' courts. The enormous difficulties of communication with England helped to foster this sense of separation. The round trip across the ocean occupied the better part of a year, and was hazardous and expensive; a colonist who had made it was a marked man—as Hawthorne said, "the petit maître of the colonies." Nor was there any very extensive exchange of ideas, for though most of the books read in the colonies came from England, the great majority of the colonists, down to the middle of the century, seem to have read little save the Bible and biblical commentaries, and in the native literature of the time one seldom comes upon any reference to the English authors who were glorifying the period of the Restoration and the reign of Anne.

196. Over half of the 1750 colonists that lived on the American seaboard had genetic links to immigrants who had arrived a century ago.

A. True
B. False
C. Can't tell

197. Which of the following statements is supported by the above passage:

A. According to Hall, America's population at the date of the Revolutionary war could be entirely traced back to 20,000 immigrants.
B. The population in the 1751 colonies was over ten times the original immigration that moved there.
C. According to Hall, in 1751 the population in the American colonies was one million.
D. According to Hall, 80,000 people led to a population of 1,000,000.

198. According to the passage, the new aristocracy that existed in the colonies was:

A. Similar to the England's.
B. Similar to European aristocratic systems in general.
C. Not based in royal governors' courts.
D. Not based on genetic lines.

199. Most of the books on board ships were Bibles and Biblical commentaries.
A. True
B. False
C. Can't tell

200. Which of these is **NOT** given as a reason for poor communications with England:

A. Travel between America and England was costly.
B. The English saw the early colonists as backwards.
C. Travel between America and England was slow.
D. Travel between America and England was dangerous.

SET 41

In discussing Russia's role in the past World War, it is customary to cite the losses sustained by the Russian Army, losses running into many millions. There is no doubt that Russia's sacrifices were great, and it is just as true that her losses were greater than those sustained by any of the other Allies. Nevertheless, these sacrifices are by far not the only standard of measurement of Russia's participation in this gigantic struggle. Russia's role must be gauged, first of all, by the efforts made by the Russian Army to blast the German war plans during the first years of the War, when neither America, nor Italy, nor Romania were among the belligerents, and the British Army was still in the process of formation.

[Secondly], and this is the main thing, the role played by the Russian Army must be considered also in this respect that the strenuous campaign waged by Russia, with her 180 millions of inhabitants, for three years against Germany, Austro-Hungary and Turkey, sapped the resources of the enemy and thereby made possible the delivery of the final blow. This weakening of the powers of the enemy by Russia was already bound at various stages of the War to facilitate correspondingly the various operations of the Allies. Therefore, at the end of the War, three years of effort on the part of Russia had devoured the enemy's forces, enabling the Allies to finally crush the enemy. The final catastrophe of the Central Powers was the direct consequence of the offensive of the Allies in 1918, but Russia made possible this collapse to a considerable degree, having effected, in common with the others, the weakening of Germany, and having consumed during the three years of strenuous fighting countless reserves, forces, and resources of the Central Powers.

Could Germany have won the War? A careful analysis of this question brings home the conviction that Germany was very close to victory, and that it required unusual straining of efforts on the part of France and Russia to prevent Germany from "winning out."

201. According to the passage, Russia's greatest contribution to the War was?
A. Contributing more sacrifices than any other ally.
B. Blasting the German war plans during the first years of the War.
C. Consuming countless reserves, forces and resources of the Germans.
D. Sapping the resources of Germany, Austro-Hungary and Turkey for 3 years.

202. How many countries were Russian allies?
A. 2.   B. 3.   C. 4.   D. 5.
E.

203. Russia was the main country fighting against Germany in the early years of the War?
A. True.   B. False.   C. Can't tell.

204. The War was won by?
A. Germany running out of resources.
B. The offensive drive of the Allies.
C. The arrival of America.
D. Turkey and Austro-Hungary changing sides.

205. If it were not for Russia, Germany would have won the war?
A. True.   B. False.   C. Can't tell.

SET 42

We freeze some moments in time. Every culture has its' frozen moments, events so important and personal that they transcend the normal flow of news.

Americans of a certain age, for example, know precisely where they were and what they were doing when they learned that President Franklin D. Roosevelt had died. Another generation has absolute clarity of John F. Kennedy's assassination. And no one who was older than a baby on 11th September, 2001, will ever forget hearing about, or seeing, aeroplanes flying into skyscrapers.

In 1945, people gathered around radios for the immediate news and stayed with the radio to hear more about their fallen leader and about the man who took his place. Newspapers printed extra editions and filled their columns with detail for days and weeks afterward. Magazines stepped back from the breaking news and offered perspective.

11th September, 2001, followed a similarly grim pattern. We watched again and again the awful events. Consumers of news learned about the attacks, thanks to the television networks that showed the horror so graphically. Then we learned some of the how's and why's, as print publications and thoughtful broadcasters worked to bring depth to events that defied mere words. Journalists did some of their finest work and made me proud to be one of them.

But something else, something profound, was happening this time around: news was being produced by regular people who had something to say and show, and not solely by the "official" news organisations that had traditionally decided how the first draft of history would look. This time, the first draft of history was being written in part, by the former audience. It was possible, it was inevitable, because of new publishing tools available on the Internet.

206. The author of this passage is a journalist.
A. True.
B. False.
C. Can't tell.

207. Which media outlet tended to share a different point of view to the others?
A. Radios.
B. Newspapers.
C. Magazines.
D. Televisions.

208. Which media outlet had the biggest effect on changing the way news is spread?
A. The Internet.
B. Newspapers.
C. Magazines.
D. Televisions.

209. Which media outlet was mainly responsible for notifying consumers of the news that events had occurred in 1945?
A. Televisions.
B. Newspapers.
C. Radios.
D. Magazines.

210. What was the big news story in 1945?
A. President Franklin D. Roosevelt died.
B. John F. Kennedy was assassinated.
C. An aeroplane exploded into skyscrapers.
D. John. F Kennedy was elected.

SET 43

There is nothing in England today with which we can compare the life of a fully enfranchised borough of the fifteenth century. The town of those earlier days, in fact, governed itself after the fashion of a little principality. Within the bounds which the mayor and citizens defined with perpetual insistence in their formal perambulation year after year, it carried on its isolated self-dependent life.

The inhabitants defended their own territory, built and maintained their walls and towers, armed their own soldiers, trained them for service and held reviews of their forces at appointed times. They elected their own rulers and officials in whatever way they chose to adopt, and distributed among officers and councillors just such powers of legislation and administration as seemed good in their eyes. They drew up formal constitutions for the government of the community, and as time brought new problems and responsibilities, more were made, re-made and revised again; their ordinances with restless and fertile ingenuity, till they had made of their constitution a various medley of fundamental doctrines and general precepts and particular rules, somewhat after the fashion of an American state of modern times.

In all concerns of trade, they exercised the widest powers, and bargained and negotiated and made laws as nations do on a grander scale today. They could covenant and confederate, buy and sell, deal and traffic after their own will; they could draw up formal treaties with other boroughs, and could admit them to or shut them out from all the privileges of their commerce; they might pass laws of protection or try experiments in free trade. Often, their authority stretched out over a wide district, and surrounding villages gathered to their markets and obeyed their laws; it might even happen in the case of a staple town that their officers controlled the main foreign trade of whole provinces.

211. In the 15th century, towns drew up treaties to defend each other in times of attack.
A. True.
B. False.
C. Can't tell.

212. How were the leaders of boroughs brought into power?
A. Power was passed down the family.
B. Invaders conquered towns and become the rulers.
C. Inhabitants elected their own rulers.
D. There was a leader of the whole country, with majors in each borough.

213. Town life in the 15th century is more comparable to the American state of modern times than England in modern times.
A. True.
B. False.
C. Can't tell.

214. The authorities in the 15th century controlled who?
A. Their respective villages only.
B. Their respective towns only.
C. Their respective towns and surrounding villages.
D. The whole country.

215. How might disputes between boroughs have been settled in the 15th century?
A. By war.
B. By banning trading between boroughs.
C. With a treaty.
D. All of the above.

SET 44

Nowhere is the influence of sex more plainly manifested than in the formulation of religious conceptions and creeds. With the rise of male power and dominion, and the corresponding repression of the natural female instincts, the principles that originally constituted the God-idea gradually gave place to a Deity better suited to the peculiar bias that had been given to the male organism. An anthropomorphic God, like that of the Jews, whose chief attributes are power and virile, could have had its origin only under a system of masculine rule.

Religion is especially liable to reflect the vagaries and weaknesses of human nature; and, as the forms and habits of thought connected with worship take a firmer hold on the mental constitution than do those belonging to any other department of human experience. Religious conceptions should be subjected to frequent and careful examination in order to perceive, if possible, the extent to which we are holding on to ideas which are unsuited to existing conditions.

In an age when every branch of inquiry is being subjected to reasonable criticism, it would seem that the origin and growth of religion should be investigated from beneath the surface and that all the facts bearing upon it should be brought forward as a contribution to our fund of general information. As well might we hope to gain a complete knowledge of human history by studying only the present aspect of society, as to expect to reach reasonable conclusions respecting the prevailing God-idea by investigating the various creeds and dogmas of existing faiths.

216. Masculine rule has always occurred.
A. True.
B. False.
C. Can't tell.

217. Which god has attributes of power?
A. All gods.
B. The Jewish god.
C. The Christian god.
D. No gods.

218.   Why is it a good idea to study the different Gods?
A. It provides an insight to societies of the past.
B. Because it is important to understand and respect other people's religion.
C. To find out which God is the most powerful.
D. To find out about the weaknesses of human nature.

219. Where is the influence of sex most clearly displayed?
A. In the formation of religion.
B. In the rules of religion.
C. By the lack of opposition against religion by women.
D. By the gods attributes.

220. The author is religious.
A. True.
B. False.
C. Can't tell.

SET 45

"What is a novel?" A novel is a marketable commodity, of the class collectively termed "luxuries," as not contributing directly to the support of life or the maintenance of health. The novel, therefore, is an intellectual artistic luxury in that it can be of no use to a man when he is at work, but may conduce to peace of mind and delectation during his hours of idleness.

Probably, no one denies that the first object of the novel is to amuse and interest the reader. But it is often said that the novel should instruct as well as afford amusement, and the "novel-with-a-purpose" is the realisation of this idea. The purpose-novel, then, proposes to serve two masters, besides procuring a reasonable amount of bread and butter for its writer and publisher, it proposes to escape from my definition of the novel in general and make itself an "intellectual moral lesson" instead of an "intellectual artistic luxury." It constitutes a violation of the unwritten contract tacitly existing between writer and reader. A man buys what purports to be a work of fiction, a romance, a novel, a story of adventure, pays his money, takes his book home, prepares to enjoy it at his ease, and discovers that he has paid a dollar for somebody's views on socialism, religion, or the divorce laws.

221. According to the author, a writer should write a novel with the sole purpose of amusement?

A. True.
B. False.
C. Can't tell.

222. Which of the following is the best use of a novel?

A. To be of use to a man when he is at work.
B. To spread the author's views on socialism, religion, or the divorce laws to those who otherwise would not listen to them.
C. To make moral lessons more interesting.
D. To pleasure a man during periods of idleness.

223. The writer of the passage regards a novel which has a strong view on socialism as their biggest hate.

A. True.
B. False.
C. Can't tell.

224. What is the unwritten contract between the author and reader?
A. A novel should provide an intellectual moral lesson.
B. A novel should have a purpose.
C. A novel should afford amusement.
D. A novel should be a work of fiction.

225. What is the most common opinion which the author does not share?
A. A novel should have a purpose.
B. A novel should instruct the reader.
C. A novel should conduce to peace of mind.
D. A novel should afford amusement.

SET 46

What is patriotism? Is it love of one's birthplace, the place of childhood's recollections and hopes, dreams and aspirations? Is it the place where, in child-like naivety, we would watch the fleeting clouds and wonder why we too could not run so swiftly? The place where we would count the milliard glittering stars, terror-stricken lest each one "an eye should be," piercing the very depths of our little souls? Is it the place where we would listen to the music of the birds and long to have wings to fly, even as they, to distant lands? Or the place where we would sit at mother's knee, enraptured by wonderful tales of great deeds and conquests? In short, is it love for the spot, every inch representing dear and precious recollections of a happy, joyous, and playful childhood?

If that were patriotism, few American men of today could be called upon to be patriotic, since the place of play has been turned into a factory, mill, and mine, while deafening sounds of machinery have replaced the music of the birds. Nor can we still hear the tales of great deeds, for the stories our mothers tell today are but those of sorrow, tears and grief. What, then, is patriotism? "Patriotism, sir, is the last resort of scoundrels," said Dr. Johnson. Leo Tolstoy, the greatest anti-patriot of our times, defines patriotism as the principle that will justify the training of wholesale murderers; a trade that requires better equipment for the exercise of man-killing than the making of such necessities of life as shoes, clothing, and houses; a trade that guarantees better returns and greater glory than that of the average workingman.

226. The author believes patriotism as being the love for the spot where one grew up and had many happy memories.
    A. True.
    B. False.
    C. Can't tell.

227. According to the author, are few American's of today patriotic?

    A. Yes, because where they grew up has been destroyed.
    B. Yes, because the current generation didn't have anywhere where they could listen to the music of the birds when they were children as the sounds of machinery has replaced this.
    C. Yes, because not many people do justify the training of wholesale murderers.
    D. No, because the definition of patriotism has not been defined and so one cannot say whether few American's of today are patriotic.

228. What is a possible negative of being patriotic?

    A. Justifying mass killing.
    B. Being depressed when one's birthplace is replaced by factories, mills and mines.
    C. Reducing production of necessities such as workplaces.
    D. Encouraging migration to distant lands.

229. The most common stories of today when children sit at their mother's knee are what?

    A. Wonderful tales of great deeds and conquests.
    B. Stories to encourage dreams and aspirations.
    C. Stories of sorrow, tears, and grief.
    D. Tales of flying away to distant lands.

230. What might a young child wonder?

    A. What distant lands are like?
    B. Whether the conquests of their Mothers stories are true.
    C. Whether their birthplace will one day be turned into a factory.
    D. Why they cannot keep up with clouds.

SET 47

The law of benefits is a difficult channel, which requires careful sailing or rude boats. It is not the office of a man to receive gifts. How dare you give them? We wish to be self-sustained. We do not quite forgive a forgiver. The hand that feeds us is in some danger of being bitten. We can receive anything from love, for that is a way of receiving it from ourselves (hence the fitness of beautiful, not useful things for a gift); but not from anyone who assumes to bestow. We sometimes hate the meat that we eat, because there seems something of degrading dependence in living by it.

He is a good man, who can receive a gift well. We are either glad or sorry at a gift, and both emotions are unbecoming. Some violence, I think, is done, some degradation borne, when I rejoice or grieve at a gift. I am sorry when my independence is invaded, or when a gift comes from such as do not know my spirit, and so the act is not supported; and if the gift pleases me overmuch, then I should be ashamed that the donor should read my heart and see that I love his commodity, and not him.

This giving is flat usurpation, and therefore, when the beneficiary is ungrateful, as all beneficiaries hate all Timons, not at all considering the value of the gift, but looking back to the greater store it was taken from, I rather sympathize with the beneficiary than with the anger of my lord, Timon. For, the expectation of gratitude is mean and is continually punished by the total insensibility of the obliged person.

231. Who is described as a good man?
A.   A man who can forgive a forgiver.
B.   A man who can give a good gift.
C.   A man who can take a gift well.
D.   A man who can enjoy the meat he eats.

232. Why do some men not like to receive a gift?
A. Because men wish to maintain themselves by independent effort.
B. They do not wish to depend upon it.
C. They only like to receive gifts of love.
D. They prefer to pick their own gifts.

233. The author of this passage does not like to receive gifts.
A. True.
B. False.
C. Can't tell.

234. When is it most acceptable to give a gift?
A. When you can read someone's heart and present the perfect gift.
B. For those most in need.
C. Never.
D. When the receiver loves you.

235. What is the major negative of enjoying a gift?
A. You have to return the favour.
B. It is going against the desire to self-sustain.
C. Appreciating the item rather than the effort of the giver.
D. All of the above.

SET 48

English writers who have spoken of Goethe's "Doctrine of Colours," have generally confined their remarks to those parts of the work in which he has undertaken to account for the colours of the prismatic spectrum, and of refraction altogether, on principles different from the received theory of Newton. The less questionable merits of the treatise consisting of a well-arranged mass of observations and experiments, many of which are important and interesting, have thus been in a great measure overlooked. The translator, aware of the opposition which the theoretical views alluded to have met with, intended at first to make a selection of such of the experiments as seem more directly applicable to the theory and practice of painting. Finding, however that the alterations this would have involved would have been incompatible with a clear and connected view of the author's statements, he preferred giving the theory itself, reflecting, at the same time, that some scientific readers may be curious to hear the author speak for himself even on the points at issue.

In reviewing the history and progress of his opinions and research, Goethe tells us that he first submitted his views to the public in two short essays entitled "Contributions to Optics." Among the circumstances which he supposes were unfavourable to him on that occasion, he mentions the choice of his title, observing that by a reference to optics he must have appeared to make pretensions to a knowledge of mathematics, a science with which he admits he was very imperfectly acquainted.

236. Mathematics was Goethe's greatest science.
   A. True.
   B. False.
   C. Can't tell.

237. What was Goethe's most popular work?

A.   His paper on 'Contribution to Optics'.
B.   His paper on 'Doctrine of Colours'.
C.   His ideas of refraction.
D.   His ideas of mathematics.

238. What was the main problem the translator faced when writing the paper 'Doctrine of colours'?

A.   He wanted to present the scientific evidence before the theory to the audience, while Goethe did not.
B.   He wanted to present the theory and then back it up with scientific evidence to the audience, while Goethe did not.
C.   He wanted to wait until there was more scientific evidence before publishing as he knew the opposition would believe Goethe's theory.
D.   The translator did not believe Goethe's theory and didn't want to write a paper which may convince others it is true.

239. How many papers did Goethe publish, according to the article?
   A. 1
   B. 2
   C. 3
   D. 4

240. What is considered Goethe's most notable work?
   A.   About colours of the prismatic spectrum and of refraction.
   B.   His contribution to optics in general.
   C.   Disproving Newton's theories.
   D.   Changing the ways in which theories are presented.

SET49

That marriage is a failure none but the very stupid will deny. One has but to glance over the statistics of divorce to realize how bitter a failed marriage really is. Nor will the stereotyped Philistine argument that the laxity of divorce laws and the growing looseness of women account for the fact that: first, every twelfth marriage ends in divorce; second, that since 1870 divorces have increased from 28 to 73 for every hundred thousand population; third, that adultery, since 1867, as ground for divorce, has increased 270.8 percent; fourth, that desertion increased 369.8 percent.

Henrik Ibsen, the hater of all social shams, was probably the first to realize this great truth. Nora leaves her husband, not as the stupid critic would have it, because she is tired of her responsibilities or feels the need of woman's rights, but because she has come to know that for eight years she had lived with a stranger and borne him children. The moral lesson instilled in the girl is not whether the man has aroused her love, but rather is it, "How much?" The important and only God of practical American life: Can the man make a living? Can he support a wife? That is the only thing that justifies marriage. Gradually, this saturates every thought of the girl; her dreams are not of moonlight and kisses, of laughter and tears; she dreams of shopping tours and bargain counters. Can there be anything more humiliating, more degrading than a life-long proximity between two strangers? No need for the woman to know anything of the man, save his income. As to the knowledge of the woman—what is there to know except that she has a pleasing appearance?

241. According to the passage, most women get married for the sake of money.
    A. True.
    B. False.
    C. Can't tell.

242. The author believes that marriages end due to the "looseness" of women.
    A. True.
    B. False.
    C. Can't tell.

243. Nora did not want children.
    A. True.
    B. False.
    C. Can't tell.

244. In America, every twelfth marriage ends in divorce.
    A. True.
    B. False.
    C. Can't tell.

245. Women always end up dreaming of moonlight kisses later in marriage.
    A. True.
    B. False.
    C. Can't tell.

SET 50

During the 1960s and 70s, terrorism was a contemporary subject due to the conflict between the UK government and the IRA. 'The Troubles' (the name given to the violence) originated in the 1920s and eventually resulted in bombings on the streets of Northern Ireland and occasionally in England. The government felt that in order to prevent mayhem, their actions needed to be swift and decisive. Thus, a series of temporary measures were initiated; policemen and soldiers all over Ulster were given the right to stop, question, search and arrest members of the public.

In 2000, the Terrorism Act 2000 was passed as a definitive measure following twenty years of temporary measures. Policemen were given wider stop and search powers and enabled to detain suspects for up to 48 hours without charge. The Act was met with strong criticism as it outlawed certain Islamic fundamentalist groups and this was seen as a portrayal of Islam as a religion that fuels terrorism. This, in turn, made it likely that discrimination would occur in the form of the disproportionate stopping and searching of Asians who were thought to 'look Muslim'.

Although, prior to the September 2001 attacks on Washington D.C, government legislation in the UK had attempted to prevent the occurrence of terrorism; the counter-terrorism strategies had focused a lot of attention on the punishment of terrorists and the criminalisation of new offences following their occurrence. However, the Anti-Terrorism Crime and Security Act 2001 (ATCSA) marked a more firm move towards the 'management of anticipatory risk' (Piazza, Walsh, 2010) which was to characterize the counter-terrorism legislation of the 21st century.

246. In the 1960s, policemen and soldiers all over Ulster were given the right to stop, question, search and arrest members of the public.
    A. True.
    B. False.
    C. Can't tell.

247. The Terrorism Act 2000 is still implemented today.
    A. True.
    B. False.
    C. Can't tell.

248. The act in 2000 was passed in an effort to combat the threat of Islamic extremists.
    A. True.
    B. False.
    C. Can't tell.

249. The act targets preventing rather than punishing terrorism acts.
    A. True.
    B. False.
    C. Can't tell.

250. Terrorism management was a big problem between 1980 and 2000.
    A. True.
    B. False.
    C. Can't tell.

Decision Making

The Basics

The Decision Making section is **brand new** to the UKCAT in 2017. It replaces an old section called Decision Analysis which involved deciphering the meaning of various coded phrases and is no longer tested.

The section lasts 31 minutes (with one additional minute to read the instructions), and there are 29 questions to be answered – so you need to work quickly and efficiently at a rate of about one question per minute. You will be presented with questions that may refer to text, charts, tables, graphs or diagrams. All of the questions are standalone and do not share data, so make sure to focus on each question independently.

This section was a component of the test last year, but was not scored. This year the section will be scored just like every other subsection, and the score from this section will contribute to your overall score and result.

The Questions

The idea behind this section is to assess how you use information and data to make decisions – a skill that is essential to working effectively as a doctor. The questions come in a variety of styles, but all are focussed on testing your decision making ability.

The questions making up the decision making section can be broken down into six main styles. All of the questions in this section will belong to one of these styles, so by familiarising yourself with the theory you will make it much easier to answer the questions. The six styles are:

1) **Logical Puzzles**
2) **Syllogisms**
3) **Interpreting Information**

4) **Recognising Assumptions**
5) **Venn Diagrams**
6) **Probabilistic Reasoning**

The questions may provide multiple pieces of information which build together to give the overall picture. Remember to consider each of these in turn and build up your understanding of the situation in pieces before bringing it all together.

The most important thing is to understand the premise of the question. There is often a clear chain of logic from the start of the question to the final answer, so the sooner you work this out, the sooner you can begin answering on the right track.

Strategy

As there are different styles of questions in this section, the best way to think about preparation is to subdivide the questions into their different styles and develop a clear approach to each

As a general principle, drawing diagrams can be helpful. If for example, the question has a number of people in it, write their names down on your whiteboard (or just the first initial to save space and time). If the question talks about compass directions, draw a simple map. If it talks about categories, a Venn diagram can help. If it involves a timetable or timed events, draw out a timeline to simplify your thought process. By using visual tools to complement the words, you make it easier to understand and solve the problem.

Logical Puzzles

The logical puzzle questions require you to make an inference based on the available information to get to the answer. Commonly, this will manifest in being presented with some background information (a general statement) and then some extra information (a specific statement), and both must be combined to make the conclusion and find the answer. Deductive reasoning can be used in either a positive way to prove something, or in a negative way to disprove something. By familiarising yourself with the structures, you will improve your ability to notice and use the relevant information.

An example of **positive inductive reasoning**:
1) All birds have wings
2) Ostriches are birds
3) Therefore the ostrich has wings

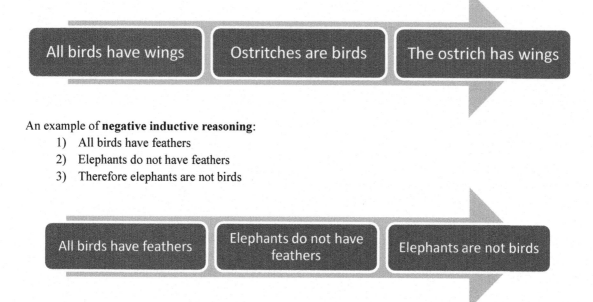

An example of **negative inductive reasoning**:
1) All birds have feathers
2) Elephants do not have feathers
3) Therefore elephants are not birds

Syllogisms

Syllogisms are an additional application of deductive reasoning. In these questions, you will be presented with information that can be used to make certain conclusions, but which will also give and incomplete description of the situation. Then, the task is to determine which answer/answers is/are supported and which aren't. To do well in these questions, you have to be very clear about the limitations of the information available – if the statement cannot be deduced from the information in the questions, then it is not true.

Interpreting information

For the interpreting information questions, you will be presented with a more complex and less directly relevant set of information than in the logical deduction questions. This information may be in the form of a passage of text describing something, or alternatively it could be in the form of a table, chart or graph. You will then have to use the information source to extract the relevant information to answer the question. Don't be afraid to use rounding and estimations – if the differences are substantial, you may not need to calculate figures exactly.

To answer these questions effectively, look at the question before digesting the data. Once you understand the premise of the question, you can approach the data in a much more focussed way to gather the information you need and ignore the distracters designed to make the question more difficult. When analysing graphs and charts, always follow a systematic approach to ensure you grasp the key message as easily as possible, such as the method illustrated below.

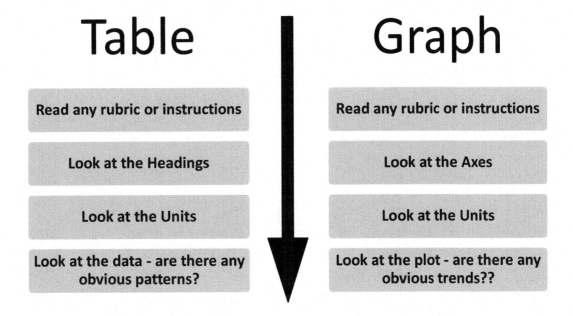

Table

- Read any rubric or instructions
- Look at the Headings
- Look at the Units
- Look at the data - are there any obvious patterns?

Graph

- Read any rubric or instructions
- Look at the Axes
- Look at the Units
- Look at the plot - are there any obvious trends??

Recognising assumptions

Assumptions are an important component of decision making, and proper use of them is essential to being a good decision maker. Relying too heavily on assumptions leaves the decision at risk of being wrong, whereas reluctance to make assumptions can make the decision process extremely slow and laborious. The questions of this style aim to probe your understanding of assumption making as a component of the decision process.

Many of these questions will ask you to select the strongest argument for or against a statement. To help you select the best option, use the acronym **FREES**. Factual – the argument should be based on fact rather than opinion. Relevant – the argument should directly address the statement in the question. Entire – the argument should address the whole question, not only one aspect of it. Emotionless – the argument should avoid emotional pleas and derive strength from relevant evidence. Sensible – the argument should be a generally sensible and reasonable approach to take.

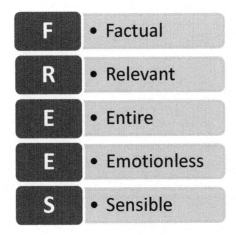

F • Factual
R • Relevant
E • Entire
E • Emotionless
S • Sensible

Venn diagrams

The category of "Venn diagram" questions encompasses any that require the use of Venn diagrams within the question. Venn diagrams might be used in the question itself to present the data, they may be required as a part of the working to deduce the correct answer, or it may be that the answers are presented in the form of Venn diagrams, and you have to choose the most appropriate response.

Venn diagrams can take a number of forms. Some may look like the style of diagram you will have seen for many years, with two or three overlapping segments into which items are sorted. But they don't necessarily look like that. Provided the basic rules are followed – that to sort any item, it is placed into each and every circle that it belongs to – then any shape can be formed. The shape and structure of the diagrams has to be altered to allow all circles that need to overlap to be able to do so. To build your skills, practice drawing Venn diagrams to classify common objects – for example vehicles, kitchen utensils, farm animals or school subjects in different ways. Below are examples of Venn diagram structures that you may encounter in this section of the UKCAT.

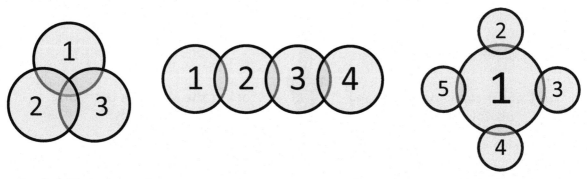

Probabilistic reasoning

These questions assess your ability to use probabilistic reasoning in the decision making process. You may be presented with probabilistic information in a variety of forms – fractions, decimals, percentages or odds – so do remind yourself of these different notations if you haven't seen them in a while. Whenever you see a probability, take the time to note exactly what occurrence the probability is representing, and whether it is the positive probability (of a thing happening) or a negative probability (of it not happening).

A good rule of thumb with probability is this. If you are asked the probability of something **or** something occurring, then the overall probability is higher than only one of them occurring so you add the probabilities. If you are asked the probability of something **and** something occurring, then the overall probability is lower than only one of them occurring, so you multiply the probabilities. For example, if asked about the probability of rain on two given consecutive days, when the probability of rain on any given day is 0.4, then it can be calculated as follows:

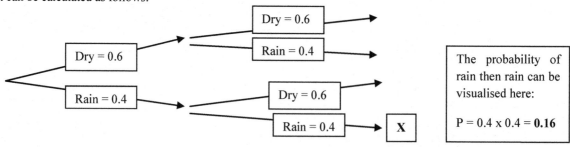

Top tip! If you are ever confused about probability, draw out a probability tree. The visualisation will enable you to see through the complexity to identify the correct calculation.

Worked examples

Example 1

A Surgeon is planning his surgery list for the next day. He has a total of 5 surgeries planned for the day. Mrs Smith has an infected arm wound that needs to be cleaned and stitched. Due to the infection, the surgeon will operate on her last.

> **Start building up the order as information in gathered**

Mr Hunt has a broken leg and will be operated on before Mr Pierce.
Mr Perry will be operated on first, his procedure will take 1hr.
Mr Dutch has his procedure 2hrs before Mr Hunt and 60mins after Mr Perry.

Mr Perry				

Which of the following statements is **TRUE**?

> **To be true, we need to see DIRECT supporting evidence**

A. Mr Hunt's leg is infected
B. Mr Dutch will be operated on third
C. There are no women on the operating list
D. **Mr Pierce will be operated on after Mr Hunt**

Answer **D**: the only order that satisfies the information is Perry-Dutch-Hunt-Pierce-Smith

Example 2

A group of scientists conduct research into the social structure of Vikings. They find that Vikings tended to be organised in close knit groups of families often living in individual villages. Leadership was often organised along performance as a leader and rarely was a birth right like in other early mediaeval societies in central Europe. Slavery was a common practice in Viking society with slaves coming from all areas raided by individual groups or from centralised slave markets.

> **This is a syllogism style – you need to work out which answers are supported by the information in the text and which ones are not. This may involve logical reasoning.**

Which of the following statements is **INCORRECT**?

A. Central European societies were controlled by birth right lordship
B. Viking society was democratic
C. The slave trade continued to be present in medieval societies
D. Viking society was close knit

Answer **B**: There is no mention of democracy; all other answers are supported by the text

> ***Top tip!*** Read the question before starting to interpret tables, charts or graphs. That way, you know what information you need and what is there to distract you!

Decision Making Questions

Question 1:

Pilbury is south of Westside, which is south of Harrington. Twotown is north of Pilbury and Crewville but not further north than Westside. Crewville is:

A. South of Westside, Pilbury and Harrington but not necessarily Twotown.
B. North of Pilbury, and Westside.
C. South of Westside and Twotown, but north of Pilbury.
D. South of Westside, Harrington and Twotown but not necessarily Pilbury.
E. South of Harrington, Westside, Twotown and Pilbury.

Question 2:

The hospital coordinator is making the rota for the ward for next week; two of Drs Evans, James and Luca must be working on weekdays, none of them on Sundays and all of them on Saturdays. Dr Evans works 4 days a week including Mondays and Fridays. Dr Luca cannot work Monday or Thursday. Only Dr James can work 4 days consecutively, but he cannot do 5.

What days does Dr James work?

A. Saturday, Sunday and Monday.
B. Monday, Tuesday, Wednesday, Thursday and Saturday.
C. Monday, Thursday Friday and Saturday.
D. Tuesday, Wednesday, Friday and Saturday.
E. Monday, Tuesday, Wednesday, Thursday and Friday.

Question 3:

If criminals, thieves and judges are represented below:

Assuming that judges must have clean record, all thieves are criminals and all those who are guilty are convicted of their crimes, which of one of the following best represents their interaction?

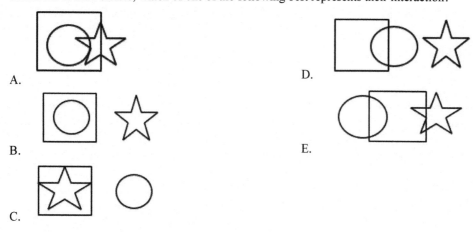

Question 4:

Apples are more expensive than pears, which are more expensive than oranges. Peaches are more expensive than oranges. Apples are less expensive than grapes.

Which two of the following must be true?

A. Grapes are less expensive than oranges.
B. Peaches may be less expensive than pears.
C. Grapes are more expensive than pears.

D. Pears and peaches are the same price.
E. Apples and peaches are the same price.

Question 5:

A class of young students has a pet spider. Deciding to play a practical joke on their teacher, one day during morning break one of the students put the spider in their teachers' desk. When first questioned by the head teacher, Mr Jones, the five students who were in the classroom during morning break all lied about what they saw. Realising that the students were all lying, Mr Jones called all 5 students back individually and, threatened with suspension, all the students told the truth. Unfortunately Mr Jones only wrote down the student's statements not whether they had been told in the truthful or lying questioning.

The students' two statements appear below:

Archie: "It wasn't Edward. "
    "It was Bella."

Charlotte: "It was Edward."
    "It wasn't Archie"

Darcy: "It was Charlotte"
    "It was Bella"

Bella: "It wasn't Charlotte."
    "It wasn't Edward."

Edward: "It was Darcy"
    "It wasn't Archie"

Who put the spider in the teacher's desk?

A. Edward
B. Bella
C. Darcy

D. Charlotte
E. More information needed.

Question 6:

On a specific day at a GP surgery 150 people visited the surgery and common complaints were recorded as a percentage of total patients. Each patient could use their appointment to discuss up to 2 complaints. 56% flu-like symptoms, 48% pain, 20% diabetes, 40% asthma or COPD, 30% high blood pressure.
Which statement must be true?

A. A minimum of 8 patients complained of pain and flu-like symptoms.
B. No more than 45 patients complained of high blood pressure and diabetes.
C. There were a maximum of 21 patients who did not complain about flu-like symptoms or high blood pressure.
D. There were actually 291 patients who visited the surgery.
E. None of the above.

Question 7:

During a GP consultation in 2015, Ms Smith tells the GP about her grandchildren. Ms Smith states that Charles is the middle grandchild and was born in 2002. In 2010, Bertie was twice the age of Adam and that in 2015 there are 5 years between Bertie and Adam. Charles and Adam are separated by 3 years.

How old are the 3 grandchildren in 2015?

A. Adam = 16, Bertie = 11, Charles = 13

B. Adam = 5, Bertie = 10, Charles = 8

C. Adam = 10, Bertie = 15, Charles = 13

D. Adam = 10, Bertie = 20, Charles = 13

E. Adam = 11, Bertie = 10, Charles = 8

F. More information needed.

Question 8:

A team of 4 builders take 12 days of 7 hours work to complete a house. The company decides to recruit 3 extra builders. How many 8 hour days will it take the new workforce to build a house?

A. 2 days

B. 6 days

C. 7 days

D. 10 days

E. 12 days

F. More information needed

Question 9:

Four young girls entered a local baking competition. Though a bit burnt, Ellen's carrot cake did not come last. The girl who baked a Madeira sponge had practiced a lot, and so came first, while Jaya came third with her entry. Aleena did better than the girl who made the Tiramisu, and the girl who made the Victoria sponge did better than Veronica.

Which **TWO** of the following were **NOT** results of the competition?

A. Veronica made a tiramisu

B. Ellen came second

C. Aleena made a Victoria sponge

D. The Victoria sponge came in 3rd place

E. The carrot cake came 3rd

Question 10:

John likes to shoot bottles off a shelf. In the first round he places 16 bottles on the shelf and knocks off 8 bottles. 3 of the knocked off bottles are damaged and can no longer be used, whilst 1 bottle is lost. He puts the undamaged bottles back on the shelf before continuing. In the second round he shoots six times and misses 50% of these shots. He damages two bottles with every shot which does not miss. 2 bottles also fall off the shelf at the end. He puts up 2 new bottles before continuing. In the final round, John misses all his shots and in frustration, knocks over gets angry and knocks over 50% of the remaining bottles.

How many bottles were left on the wall after the final round?

A. 2

B. 3

C. 4

D. 5

E. 6

F. More information needed

Question 11:

A bus takes 24 minutes to travel from White City to Hammersmith with no stops. Each time the bus stops to pick up and/or drop off passengers, it takes approximately 90 seconds. This morning, the bus picked up passengers from 5 stops, and dropped off passengers at 7 stops. What is the minimum journey time from White City to Hammersmith this morning?

A. 28 minutes

B. 34 minutes

C. 34.5 minutes

D. 36 minutes

E. 37.5 minutes

F. 42 minutes

Question 12:

I look at the clock on my bedside table, and I see the following digits:

However, I also see that there is a glass of water between me and the clock, which is in front of 2 adjacent figures. I know that this means these 2 figures will appear reversed. For example, 10 would appear as 01, and 20 would appear as 05 (as 5 on a digital clock is a reversed image of a 2). Some numbers, such as 3, cannot appear reversed because there are no numbers which look like the reverse of 3.

Which of the following could be the actual time?

A. 15:52   D. 12:22
B. 21:25   E. 21:52
C. 12:55

Question 13:

Ryan is cooking breakfast for several guests at his hotel. He is frying most of the items using the same large frying pan, to get as much food prepared in as little time as possible. Ryan is cooking Bacon, Sausages, and eggs in this pan. He calculates how much room is taken up in the pan by each item. He calculates the following:

- Each rasher of bacon takes up 7% of the available space in the pan
- Each sausage takes up 3% of the available space in the pan.
- Each egg takes up 12% of the available space in the pan.

Ryan is cooking 2 rashers of bacon, 4 sausages and 1 egg for each guest. He decides to cook all the food for each guest at the same time, rather than cooking all of each item at once.

How many guests can he cook for at once?

A. 1   D. 4
B. 2   E. 5
C. 3

Question 14:

Northern Line trains arrive into Kings Cross station every 8 minutes, Piccadilly Line trains every 5 minutes and Victoria Line trains every 2 minutes. If trains from all 3 lines arrived into the station exactly 15 minutes ago, how long will it be before they do so again?

A. 24 minutes   C. 40 minutes   E. 65 minutes
B. 25 minutes   D. 60 minutes   F. 80 minutes

Question 15:

In how many different positions can you place an additional tile to make a straight line of 3 tiles?

A. 6   D. 9
B. 7   E. 10
C. 8   F. 11

Question 16:

Ellie, her brother Tom, her sister Georgia, her mum and her dad line up in height order from shortest to tallest for a family photograph. Ellie is shorter than her dad but taller than her mum. Georgia is shorter than both her parents. Tom is taller than both his parents. If 1 is shortest and 5 is tallest, what position is Ellie in the line?

A. 1
B. 2
C. 3

D. 4
E. 5

Question 17:

Miss Briggs is trying to arrange the 5 students in her class into a seating plan. Ashley must sit on the front row because she has poor eyesight. Danielle disrupts anyone she sits next to apart from Caitlin, so she must sit next to Caitlin and no-one else. Bella needs to have a teaching assistant sat next to her. The teaching assistant must be sat on the left hand side of the row, near to the teacher. Emily does not get on with Bella, so they need to be sat apart from one another. The teacher has 2 tables which each sit 3 people, which are arranged 1 behind the other. Who is sitting in the front right seat?

A. Ashley
B. Bella
C. Caitlin

D. Danielle
E. Emily

Question 18:

Piyanga writes a coded message for Nishita. Each letter of the original message is coded as a letter a specific number of characters further on in the alphabet (the specific number is the same for all letters). Piyanga's coded message includes the word "PJVN". What could the original word say?

A. CAME
B. DAME
C. FAME

D. GAME
E. LAME

Question 19:

Lauren, Amy and Chloe live in different cities across England. They decide to meet up together in London and have a meal together. Lauren departs from Southampton at 2:30pm, and arrives in London at 4pm. Amy's journey lasts twice as long as Lauren's journey and she arrives in London at 4:15pm. Chloe departs from Sheffield at 1:30pm, and her journey lasts an hour longer than Lauren's journey.

Which of the following statements is definitely true?

A. Chloe's journey took the longest time.
B. Amy departed after Lauren.
C. Chloe arrived last.

D. Everybody travelled by train.
E. Amy departed before Chloe.

Question 20:

Jina is playing darts. A dartboard is composed of equal segments, numbered from 1 to 20. She takes three throws, and each of the darts lands in a numbered segment. None land in the centre or in double or triple sections. What is the probability that her total score with the three darts is odd?

A. $\frac{1}{4}$
B. $\frac{1}{3}$
C. $\frac{1}{2}$

D. $\frac{3}{5}$
E. $\frac{2}{3}$

Question 21

Should nurses be encouraged to make more significant decisions regarding treatment in the health care setting?

Select the strongest argument from the statements below.

A. No. Nurses are not able to make complex decisions that require a higher degree of training.
B. No. Nurses are not able to understand the complexity of medical care.
C. Yes. Nurses interact with the patients every day and therefore will be more qualified to make decisions than doctors.
D. Yes. Nurses provide valuable input to care delivery already as they deliver an important additional perspective.

Question 22

On a walk in the woods, James observes a variety of wildlife species. He finds that birds tend to fly off as soon as he approaches their position, whereas the many squirrels he sees tend to wait and observe his behaviour for some time. He also manages to see some deer in the distance and something he thinks was a fox.

Which of the following statements are true?

A. Squirrels are uncommon in the woods.
B. Birds are used to human presence.
C. He must be walking far away from any towns.
D. None of the above.

Question 23

"The physician must be able to tell the antecedents, know the present, and foretell the future – must mediate these things, and have two special objects in view with regard to disease, namely, to do good or to do no harm" – Hippocrates.
Which of the following statements is true with regards to the above statement?

A. A physician must be all-knowing and not make any mistakes.
B. It does not matter how well trained a physician is.
C. Avoidance of harm is one of the guiding principles of medicine.
D. Techniques of the past are still the best today.

Question 24

One of the biggest challenges facing the NHS is the discharge of patients from hospitals. This can generally have a variety of reasons, but most commonly it is due to the lack of adequate care being available for the patient outside of the hospital. The cuts to social care spending as well as a decreasing ability of families to care for their relatives compound this problem.
Which of the following represent valid solutions to this problem?

A. Discharge patients irrespective of availability of social care.
B. Force families to care for their patients.
C. Charge wealthy patients for extended hospital stays.
D. None of the above.

Question 25

A farmer has a 200 head herd of animals consisting of cows, goats and pigs. The total value of the herd is £50,000. Every goat is worth £30 and every pig is worth £70. Due to local regulations, he must always have twice as many goats and twice as many pigs than he has cows. How many cows does the farmer have?

A. 100
B. 40
C. 60
D. 75

Question 26

In study of wild monkeys in the Amazon rainforest, a scientist finds that the monkeys mingle according to a complex set of rules based on their gender and age. He finds that young animals mingle with others irrespective of age and gender; adolescent males mingle only with young animals and adolescent females mingle exclusively with older females and the young monkeys. The majority of older males exhibit hostile behaviour to adolescent males. Older females treat adolescent males with suspicion but tolerate them in their proximity.

Which of the following statements are true with regards to the interactions of the group?

A. Young apes never mingle with older male apes.
B. Adolescent males have the widest range of social interactions in the herd.
C. This study is irrelevant as it is not complex enough.
D. There is a strict separation between adolescent males and females.

Question 27

Steve is going hiking with a friend. In the run up to the trip, they split up the equipment. Steve takes the tent and the provisions whilst his friend carries the sleeping bags, ground mats and cooking equipment.

Place yes or no depending on if the conclusion is correct.

A. Steve's friend is lazy as he lets you carry the majority of the equipment.
B. The two depend on each other for a comfortable trip.
C. The friend must be weaker as their load is lighter.
D. Steve and his friend are likely to spend several days on their hiking trip.

Question 28

Anna is visiting a concert with her four friends, Louise, Maria, Sophie and Jenny. The five manage to get a row of seats right in front of the stage.

Anna is the tallest and stands right in the middle next to Louise.
Sophie is left handed and prefers to sit on the far left.
Jenny met another friend and sits at the far end of the row next to Louise.

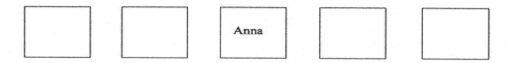

Please indicate which statements can be concluded to be true

A. Maria sits between Sophie and Anna.
B. Sophie sits at the edge of the concert hall.
C. Jenny left the group and sits with her other friend.
D. Louise sits on the second seat from the right.

Question 29

Sven is trying to set up a vegetable garden. He is allergic to tomatoes so will not be planting these. He decides to plant several rows of salad, carrots and courgettes. He also decides to plant raspberries and blackberries, even though they are not vegetables.

Which of the following conclusions are true?

A. Sven will grow more than 2 types of vegetables.
B. The garden will contain only green vegetables.
C. Allergies to tomatoes are very common.
D. None of the above.

Question 30

Louise is organizing a trip to Europe for her friends. They plan to leave London in June and visit 4 cities, starting the furthest north and move further south to then spend a week at the beach. They plan to visit Berlin, Amsterdam, Paris and Rome in that order.

Which of the following conclusions follow?

A. Paris is the most Southern city they visit.
B. Amsterdam is further north than Berlin.
C. They will be in at least 5 different countries in June.
D. None of the above

Question 31

James is planning to invest into a company by buying shares. He decides that he does not want to invest into any company that may take part in military supply. He finds that 40% of electronics firms contribute to military applications, but limited exclusively to computer technology firms. He also finds that all British steel companies are involved in producing specialty steel for the arms industry. The food industry, with exception to the beef industry does not seem to have any military contracts.

Decide if the statement is true or false?

A. James is free to invest in computer technology.
B. Investments into the chocolate industry are possible.
C. Most branches of industry are connected to military applications.
D. Investments in general are not a good idea due to the volatile political situation at the moment.

Question 32

A group of 5 friends live on the same road in a row of houses running parallel to the road.

> Austin lives the furthest from Mark.
> Steve lives two houses down from Austin.
> David lives next to Mark.

Which of the following is true?

A. Steve lives equidistant from Austin and Mark.
B. The five live in a small town.
C. Peter lives next to Austin.
D. Peter and David are neighbours.

Question 33

A group of scientists investigate male pattern baldness for a popular shampoo producer that aims to release a product for that market. They find that male pattern baldness is becoming increasingly common in men aged between 30 – 60 and that it sometimes is associated with the use of anabolic steroids or type 2 diabetes. They also find that in the majority of cases there seems to be a genetic predisposition of unknown origin, but independent of external factors.

Which of the following is are reasonable conclusions from the above?

A. Baldness is not exclusive to men.
B. There is a connection between life-style choices and hair loss
C. A hair loss treatment in shampoo form promises to help for the majority of cases.
D. Anabolic steroids can cause hair loss.

Question 34

A farmer has a forest planted. It contains exclusively needle trees, except for some oaks and beeches. It also contains fruit trees.

Which of the following is true?

A. The forest will only deliver lumber.
B. There are only 2 types of trees in the forest.
C. One can harvest Christmas trees from the forest.
D. The forest covers a large area.

Question 35

Penguins live in large colonies in various climates of the Southern hemisphere. These colonies provide protection for the individual as well as offering companionship and breeding partners. Penguins are flightless but are excellent swimmers and spend about half of their life in the ocean. They live mostly of sea animals such as fish, krill and squid.

Which of the conclusions are false?

A. Penguins spend 50% of their life on land.
B. Penguins are herd animals.
C. Penguins live mostly off seafood.
D. Penguins live exclusively on ice.

Question 36

Should people stop burning fossil fuels immediately?

Select the strongest argument from the statements below.

A. Yes, because they represent old technology that has since been overcome.
B. Yes, there is little oil and coal in the UK and therefore fossil fuels are economically unsound.
C. No, fossil fuels represent a safe and plentiful source of energy.
D. No, we are too dependent on energy to be able to source adequate supply from non-fossil sources.

Question 37

In a national election 60% of the population go to vote. Some people did not like the candidates of any party.

Does this mean that 40% of people decided not to vote? Please choose the most appropriate answer

  A. Yes, because the total would equate to 100%.
  B. Yes, because there was discontent with the candidates up for election.
  C. No, because not the entire population is allowed to vote.
  D. No, because voting is more about intuition than actual decisions.

Question 38

With increasing parental age, the risk of congenital abnormalities in babies increases. This is thought to be due to a variety of factors including accumulation of genetic defects in egg cells as well as an increased degradation of sperm quality.

Which of the following consequences is true?

  A. Older parents are less likely to have children with chromosomal abnormalities.
  B. From a genetic perspective, it is safer to have children at a younger age.
  C. Congenital abnormalities are due to poor sperm quality as sperm cells are present at birth and exposed to environmental mutagens.
  D. None of the above

Question 39

5 drivers take part in a car race. The top three drivers are each separated by 1 second. The fastest driver, James, reached the finish line 10 seconds before the last, Lucas. Felix was 5 seconds faster than Lucas, but 3 seconds slower than Peter. Dorian finishes second.

Which of the following is true?

  A. Peter finished 8 seconds before Lucas.
  B. James finishes 2 seconds before Dorian.
  C. Dorian finishes third.
  D. None of the above.

				Lucas

Question 40

A group of friends buys an assortment of junk food. They buy biscuits, chocolate and gummy bears. All but 2 of them like gummy bears. 6 like chocolate and 3 like biscuits. 1 doesn't like any sweets and therefore bought peanuts. Out of the 5 that like gummy bears, 2 like chocolate and 1 likes biscuits. 2 of the 3 that like biscuits also like chocolate.

Which of the following best represents the group's food preferences?

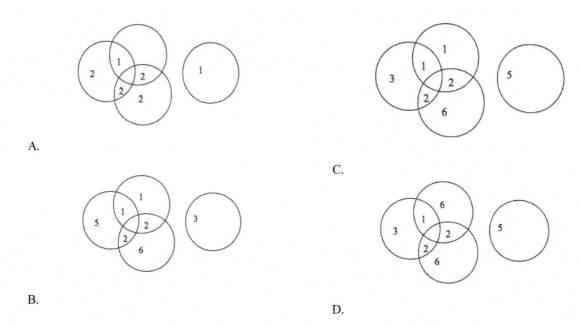

A.

B.

C.

D.

Question 41

A popular confectionary manufacturer analyses their yearly sales. They have 5 main products. All numbers are given in 1 ton batches and each batch has the same value. Their first product has sold 10 000 batches. That is 10% more than the previous year. Product 2 has been sold in 500 batches, which represents a 95% drop in sales since the last year. Product 3 has been sold 12 000 times, representing an increase of 30% from last year. Products 4 and 5 have been sold 5000 times which is the same amount as the previous year in both cases.

Which of the following conclusions must be correct?

A. The company has made more money than the previous year.
B. The company has grown internationally.
C. The company has made less money than last year.
D. None of the above.

Question 42

A plastic surgeon claims that the majority of procedures he does fall into 4 main categories, listed according to prevalence: liposuction, nose reconstruction, facial lifting and eye surgery. They all have roughly equal durations, but the price of the surgeries varies. A nose reconstruction costs roughly £500, liposuction costs £300, facial lifting costs £700 and eye surgery costs £900. The prices are graded according to clinical difficulty.

Which of the following is not a valid conclusion?

A. The surgeon does only private work.
B. The surgeon does more liposuction that eye surgeries.
C. Eye surgery is the most challenging procedure.
D. Nose reconstruction is more difficult than liposuction but less difficult than facial lifting.

Question 43

In a group of 4 friends, each has a 50% chance that they are currently smoking. Given at least two of them are currently non-smokers.

Is the probability that there are 2 smokers 50%?

A. Yes, since there is a 50% chance of one of the friends smoking.
B. Yes, since two friends are non-smokers.
C. No, since any one of them might have smoked in the past.
D. No, since the chance of both of them being smokers is 25%.

Question 44

Studies show that children from families where both parents have university degrees are twice as likely to successfully complete a university degree themselves than children from families where only one of the two has a university degree. These families in turn are twice as likely to complete a university degree than children from families where neither parent has a university degree.

Which one of the following conclusions must be correct?

A. The education system must be reformed.
B. Families where only one of the parents have university degrees are more likely to be wealthy.
C. Universities are used to maintain old relationships of master and servant.
D. None of the above.

Question 45

On a particularly busy night in an accident and emergency department in a London hospital, the staff members conduct a study into the reason of visit. The study identifies 5 main reasons for patients to seek help.

A presentation due to respiratory causes was twice as likely as a presentation due to cardiovascular causes.
The chance of being admitted due to abdominal disease was 15%, making it proportionally 10% less common than cardiovascular causes.
Traffic accidents and work accidents were equally as common.

Which of the following is true?

A. 5% of cases were due to traffic accidents.
B. 50% of cases were due to cardiovascular reasons.
C. The likelihood of a patient presenting with respiratory problems was 15%.
D. None of the above.

Question 46

A paediatric unit conducts a study into the prevalence of allergies in the children that are admitted to the unit. They find that 15% of children have allergies, with the majority of them, 65%, being food intolerances such as nut allergies. This was followed by contact allergies to latex, accounting for 10% of allergies. Other common allergies included dust and pollen.

Which of the following conclusions are correct?

A. Allergies are a problem affecting a large proportion of children.
B. Roughly 10% of all children have food allergies.
C. Nuts are the most common reason for allergies.
D. Dust and pollen allergies are much more common in adults.

Question 47

In the wild, chimpanzees have been observed to be using simple tools for food gathering as well as for hunting. It is not uncommon to see them fishing with sticks for algae or see them use rocks for opening nuts and seeds. In particular, over the last two decades, this has become an increasingly common observation.

Which of the following conclusions cannot be reasonably drawn?

A. Chimpanzees are becoming more intelligent.
B. Chimpanzees understand the effect of tools.
C. Chimpanzees are able to manipulate their environment.
D. None of the above.

Question 48

Carl and his friends want to build a treehouse. They know that they will need strong material as all 5 of them are supposed to be able to be on the treehouse at the same time. They calculate that the platform must be able to hold 225kg of weight for all of them as well as the structure of the house itself.

Carl is 10kg heavier than Alex.
Peter is the lightest of them, weighing 35kg.
Ben is the same weight as Alex
Luke weighs 5kg more than Alex and 10kg more than Peter.

Which of the following must be true?

A. Carl is neither the heaviest nor the lightest.
B. Ben weighs 10kg less than Carl.
C. The house can weigh up to 25kg.
D. Luke is the heaviest.

Question 49

A group of scientists study different subspecies of a spider family. They find that some of the subspecies share traits with one another, whilst others have nothing in common.

Group A shares some traits with group C.
Group C shares traits with group D but not with Group B.
Group C shares traits with A.
Group E shares traits with groups A.
Group E shares traits with group C but with neither Group A nor Group B.

Which of the following represents the interrelation between the groups the best?

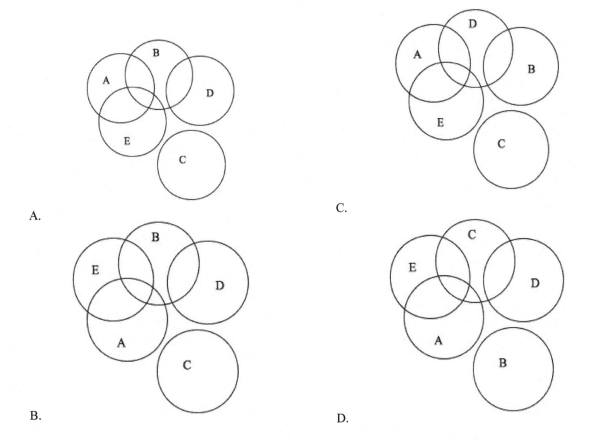

A.

B.

C.

D.

Question 50

Ant colonies are organised as large units with clear distribution of responsibility amongst individual groups of individuals, coordinated by communication mechanisms. They are headed by a queen and thereby represent a female dominant society. Some species of ants feed of plant material, other species feed of animal material. Multiple colonies can exist in proximity leading to intraspecies competition but little interspecies competition.

Which of the following must be true?

A. Ants compete amongst individuals for food.
B. Ants of different colonies, but of the same species work together to displace ants from different species.
C. Ant colonies exist based on distribution of labour.
D. Ants are exclusively herbivores.

Question 51

Should the use of pesticides be outlawed by the government? Select the strongest argument from the list below.

A. Yes, pesticides should be outlawed as they are unnatural.
B. No, pesticides should not be outlawed as they allow the UK food industry to remain internationally competitive.
C. Yes, pesticides should be outlawed as they carry significant health risks for the consumer.
D. No, pesticides should not be outlawed as they represent an important branch of industry employing many workers

Question 52

Peter loves pink marbles. They represent the majority of marbles he owns. His sister likes marbles too. She has mainly blue ones but also some yellow and green ones. She trades him some of these for some of his pink ones. Now Peter also has a few yellow and green marbles, but he does not like them as much.

Select true or false for the conclusions below

A. Peter likes blue marbles.
B. Peter has less green marbles than pink ones.
C. Peter has no yellow marbles.
D. Peter should not play with marbles as they are a choking hazard.

Question 53

Ancient peoples in hunter gatherer societies lived a nomadic lifestyle, roaming freely through the forests of Europe. Before developing the technology needed for cultivation of grain, the majority of their food consisted of meat as this was available year around. After developing farming and setting up settlements, the proportion of non-meat food materials in their diet increased to include grain and other produce as well as by-products of animal farming.

Which of the following conclusions are false?

A. Hunting represented an important food source for our ancestors.
B. Farming formed the basis of settlement development.
C. Nomadic people tend to have a lower proportion of farm produced foods in their diet.
D. None of the above.

Question 54

Different studies have demonstrated that physical activity in children increases their cognitive performance.

Which one of the following conclusions can be drawn from this?

A. Exercise should play a bigger role in school life.
B. Leading a sedentary life style can have detrimental effects on intellectual development of children.
C. Children that are members in sports clubs are likely to perform better at school.
D. All of the above.

Question 55

Recently NASA discovered several planets that stand in a similar constellation to their sun as our planet does with our sun. This led to general conclusion that there are further inhabitable planets in the universe.

Which of the following assumptions must be met in order for this conclusion to be true?

A. The new planets must meet other pre-requisites for life such as air and water.
B. The planets are already inhabited.
C. The sun the new planets are related to must be stronger than ours.
D. None of the above.

Question 56

Different birds have different food preferences. Species A mainly eats sunflower seeds, but if that is not available may fall back to pumpkin seeds. Species B has grown particularly strong jaws allowing them to break open the hard shells of hazelnuts which represent their favoured food source. Species C mainly lives off berries, but will also eat the smaller seeds or insects if berries are not available. Species D lives off smaller insects.

Which of the following is true?

A. Species B lives off berries.
B. Species D has a strong beak to kill its insect pray.
C. Species A exclusively eats pumpkin seeds.
D. Species C is adapted to all seasons.

Question 57

NHS employees should be encouraged not to smoke, drink or be overweight as they represent important role-models to society.

Which represents the biggest problem with this statement?

A. Healthcare professionals are not role-models for society.
B. Politicians are more important in guiding public opinion.
C. This statement limits freedom of choice.
D. Drinking, smoking and obesity represent important health risks.

Question 58

A group of scientists conducted a study into the purchasing behavior of young men. The find that having positive experiences such as holidays has the greatest effect in encouraging purchase, whereas advertising depicting products themselves without a theme has the least effect on encouraging purchase. Other effective techniques to encourage purchase seem to be financial success or social interaction such as with friends.

Which of the following statements are correct?

A. Financial gain is powerful advertising tool.
B. Advertising is most effective if exclusively focused on the product itself.
C. Young women are most encouraged to make purchases by positive experiences.
D. None of the above.

Question 59

A company groups applicants into different populations depending on the type of experiences they bring to the table.

Most applicants posses good excel skills, but only few have good photoshop skills.
Some of those that have good photoshop skills, speak more than one language.
A proportion of those that speak more than one language also have excellent communication skills.
Very few applicants speak only English, but have great experience from previous jobs.

Which of the following represents the distribution of applicants the most accurately?

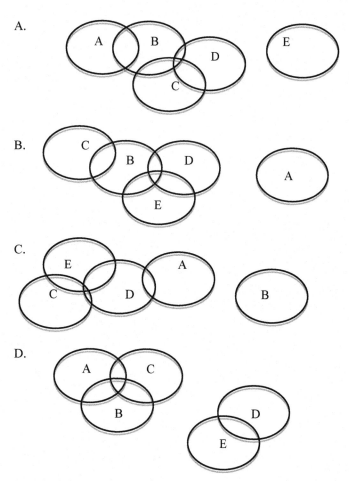

A.

B.

C.

D.

Question 60

A group of five friends compare the results of a recent maths test. Tom finds he achieved the highest score in all three sections. Richard has the second best score in sections A and B but the worst score in section C. Tony has the same result as Richard in section C, and worse results than Anthony in sections A and B. Anthony is in the middle field achieving 50% of all possible points in all three sections. Steve finished better than Richard in section C, but worse than Anthony in sections A and B.

Which of the following is false?

A. Tom has the best results.
B. Steve has a higher score than Tony.
C. Anthony has worse results for section A and B than Richard.
D. Richard finishes well in section C.

Question 61

A tobacco company orders research into the effects of a new cigarette that is supposed to be less toxic than the current most common cigarettes on the market. The results show that the new cigarette has a lower tar content than competitor brands, but the amount of nicotine is twice as high and so is the amount of carbon monoxide that's inhaled. They also find that there is significantly more nitrous oxide released from the new cigarette than from competition blends.

Which of the following is correct?

A. The new cigarette leads to less carbon monoxide being inhaled.
B. The new cigarette contains more tar being inhaled.
C. The competitors blend had to be smoked in twice the amount to reach the same nicotine level.
D. Overall the new cigarette blend is associated with less toxicity.

Question 62

A train travels from Elmsworth to Southwarf. The trip takes a total of 120mins. After 30mins of travel the train reaches the Eastwich where it stops for 5mins. After that it travels for another 45mins to reach Northtown. From Northtown the trip to Southwarf takes another 30mins, but the train waits for 5mins at the station before setting off on the journey.

Which of the following is correct?

A. The trip from Elmsworth to Northtown takes 70mins.
B. Eastwich lies 60mins away from Southwarf.
C. The train waited for 5mins at Northtown.
D. All of the above.

Question 63

A sports club publishes the results of a local football tournament. There are a total of 5 teams taking part but only the first 2 teams will receive a prize. In the scoring system, a victory will deliver 3 points, a loss will deliver 0 and a draw will deliver 1 point to each team. In the tournament each team plays each other team once.

Team A wins against team E and C, but draws against team B and loses to team D.
Team B never wins but finishes with 3 points.
Team C wins against Team E and draws with Team B.
Team D does not lose or draw any games.

Which of the following are correct?

A. Team D will finish with the second most points.
B. Team E draws with team B.
C. Team C will finish first.
D. It is customary for the best 3 teams to receive a prize.

Question 64

Every year there is a discussion surrounding the discontinuation of daylight saving time.

Which of the following represents the strongest argument in support of this?

A. Daylight saving is not necessary as it was introduced during the resource shortage of World War 1 and now this war is long over.
B. It is confusing to citizens who forget to change their watches.
C. It causes Jetlag.
D. Especially in the early months of spring it places the end of school in a dangerous twilight time causing increased casualties to road traffic accidents.

Question 65
Tim plans to run a marathon. For that he asks his friends to place themselves at different intervals from the start line to hand him snacks and sports drinks. He places his friend Chris at 15km after the start line. Anne stands 31km after the start line followed by Tara 5km after her. His friend Peter is placed 2km before the finish. Philip stands 10km up race from Chris.

Which of the following is true?

A. Chris stands 10km from Anne.
B. Philip is the first friend Tim will pass.
C. The distance between snack stops decreases towards the end of the race .
D. All of the above.

Question 66
"When conducting research, all funding with corporate interest should be refused."

Select the strongest argument supporting this statement.

A. The government should fund all research.
B. Corporations never deliver enough funding.
C. Research should not be about the money.
D. Corporate funding can restrict impartial research.

Question 67
Lisa likes to put her 5 dolls in a specific order when she puts them away on her shelf at the end of the day.

Tina always sits next to Annette.
Annette sits the furthest from Stephanie.
Patricia sits next to Tina.

Stephanie				

Which of the following is correct?

A. Patricia sits next to Annette.
B. Tina must sit at the edge.
C. Sophie must sit next to Stephanie.
D. Patricia does not sit in the middle.

Question 68
When it comes to the actions of the individual, there are several priorities influencing behaviour. Factors like socialization and inter-personal relationships can influence this decision making as well as some degree of individual variability. It is for this reason that we can find all kinds of behavioral patterns in people from all different walks of life, independent of the cultural background, but they all follow similar patterns.

Which of the following conclusions can be drawn from the above?

A. People make choices based on fixed patterns that show little variability.
B. Certain behavioral traits are associated with specific social classes.
C. Poor people will always act differently than rich people.
D. None of the above.

Question 69
Steve ranks his 5 university applications.

He really wants to go to Cambridge.
He ranks Sheffield in third place, behind Imperial.
He places Southampton between Cardiff and Sheffield.

Cambridge				

Which of the following is correct?

A. Cardiff must be his last choice.
B. Sheffield is higher up in his ranking than Imperial.
C. His first choice is Oxford.
D. All his universities are private schools.

Question 70
If we don't change our wasteful, consumerist behavior, we as a species will fail.

Which of the following represents the strongest argument in support of this question?

A. Consumerism is bad.
B. Due to the increasing shortage of resources, we must start to conserve resources more.
C. Capitalism is based on consumerism.
D. Our current behavior is damaging to the Earth upon which we rely for sustenance.

Question 71
Tim is going to the dentist. He hates going to the dentist. In order to make the visit easier for him, he tries to distract himself with little games that keep his mind occupied and prevent him from focusing on what is going on around him. His favorite coping mechanism is reciting poetry in his head. He also likes to tell himself stories that distract him.

Which of the following is true?

A. Tom hates dentists.
B. Tim likes poetry.
C. Coping mechanisms are the best way to deal with stressful situations.
D. All of the above.

Question 72
A group of scientists research swarm patterns of small fish. They find that there is a great degree of coordination between the individual fishes in the swarm in order to maintain a tight group. They also find that swarm patterns seem to be particularly successful in more murky water in comparison to very clear water. There is also a difference in success rate depending on the eyesight of the number one predator in the area, worse eye sight seems to increase protection to the individual.

Which of the following conclusions can be drawn?

A. Large fish never swim in swarm patterns.
B. The water temperature is a central factor in swarm formation.
C. The better the eyesight of the predator, the more success swarm patterns show in protecting the individual.
D. There seems to be a visual component to the success of swarm formations in protecting the individual.

Question 73
A political party is trying to determine the most popular topics in order to prepare their manifesto for the next general election. They produce a survey asking people to select the topic they find most pressing. 50% of people consider health care to be essential. Defense spending is considered the most essential by only 10% of people filling in the survey. Education comes in as the second highest priority, accounting for 25%. The remaining people consider infrastructure to be the most relevant.

Which of the following is correct?

A. Healthcare is not an issue in this population.
B. The population must feel very secure.
C. The study is flawed as it has too few choices.
D. Infrastructure is an important issue for 15% of people.

Question 74
A local political leader conducts a survey into the key interests of his constituents. He finds that the issues can be reduced to five main issues: Education, housing, health care, smoking policy and public cleanliness. Education is the most important point accounting for 28% of votes, followed by housing which is most important to 5% less of the voters. The least important issue is smoking policy, a priority for only 11% of people, which places it behind public cleanliness policy which represents 16% of all votes.

Which of the following conclusions is incorrect?

A. Health care is more important than public cleanliness policy.
B. Housing is more important than public health.
C. Smoking is the least important issue.
D. Housing policy accounts for 22% of votes.

Question 75
The mating behavior of birds heavily relies on the use of loud calls that are designed to attract the attention of a female to the male where the loud call originates. They are species specific, though there can also be a certain degree of mimicry when it comes to breeding calls. The amount of different calls varies depending on the season and increases with the amount of birds that are actively looking for a sexual partner.

Which of the following conclusions can be drawn from the above text?

A. It never happens that one bird uses another species' mating call.
B. Birds are the most active in the summer.
C. Females sing to attract males.
D. None of the above.

Question 76
Elephants are highly social animals. They live in groups that are led by a single older female elephant, and the main aim of the group is to protect the young whilst providing the most stable social environment for the young to grow up. For this reason, young males are often excluded from the groups in order to prevent conflict for hierarchical reasons and to encourage the young males to find their own herd, thereby also contributing to genetic variability by preventing inbreeding.

Which of the following statements about elephant behavior is correct?

A. Elephants live in group led by males.
B. Elephant groups include young animals that are protected by the group.
C. Female elephants share the responsibility of leading the group.
D. All of the above.

Question 77

A group of monkeys is found to prefer yellow fruit to all other types of fruit. They will also eat red fruit. They never eat green fruit.

Which of the following statements is true?

A. The monkeys prefer red fruit to green fruit.
B. The monkeys eat four different colored fruits.
C. As the monkeys don't eat the green fruits, they must be poisonous.
D. The statement is flawed as the data is too limited.

Question 78

A group is only as strong as its weakest member.

Which of the following statements most supports this theory?

A. Groups consist of individuals, so without them, there is no group.
B. A large group is better than a smaller one.
C. Groups are based on division of labour, therefore struggling members will limit efficiency.
D. All of the above.

Question 79

A mobile phone company takes stock of the sales of their 5 different handsets. In total they sold 10,000 handsets in the last year.

Set A sold the most, accounting for 4,000 units sold.
Set B and C sold to equal shares.
Set D sold 1,500 units.
Set E sold worst, with only 500 units sold.

		Set C		

Which of the following statements is true?

A. Set C sold 2,000 units.
B. Set B sold 2,500 units.
C. Set E must be the most expensive.
D. Set D is the second most sold handset.

Question 80

Different types of trees prefer different conditions. A study delivers the following results:

Walnut trees like moist ground but secrete a poison that prevents any other trees from growing around them.
Willow trees like darker and moist areas.
Beeches grow high and grow in pretty much all areas.
Oaks dislike moist areas.

Which of the following best represents the individual relationships?

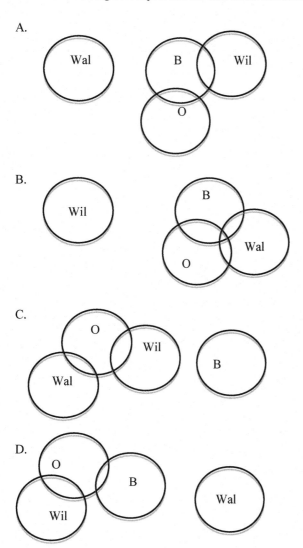

A.

B.

C.

D.

Question 81

Five friends decide to order take-away. They decide to order Chinese food, as 80% of the group are happy with Chinese. They order three portions of egg fried rice, two portions of beef spring rolls, 2 portions of sweet and sour chicken and one portion of sweet and sour duck. They also order two fish dishes.

Which of the following conclusions can be reasonably drawn?

A. 4 of the friends wanted Chinese food.
B. There are vegans in the group.
C. One friend wanted Pizza instead of Chinese.
D. One of the friends must have an allergy to eggs as they only order three portion of egg fried rice.

Question 82

After a local sports tournament, the leader table of countries is published, based on gold medals achieved. All different groups have the same number of athletes and there is a total of 5 groups.

Group A achieves 5 gold medals, 7 silver medals and no bronze medals.
Group B achieves 6 gold medals, 2 silver medals and 1 bronze medal.
Group C Ranks highest.
Group D achieves 1 gold medal and 10 silver medals as well 3 bronze medals.
Group E achieves 3 bronze medals.

Which of the following is correct?

A. Group A ranks higher than group B.
B. Group C has achieved at least 7 gold medals.
C. Group D finishes last.
D. Group C has more Bronze medals than group D.

Question 83

Scientists have found that due to the change in temperatures during the winter, the behavior of birds that normally fly South for warmer climates is changing. They now are more likely to stay in their normal habitat without migrating. This results in a constant presence of the birds which influences predator as well as prey populations. Feeding habits of humans in developed countries have however helped to counterbalance the increased pressure on prey populations.

Which of the following conclusions cannot be drawn from the text?

A. Migratory birds provide important prey to predators.
B. Climate change has reduced bird migration
C. It must be getting colder in the South which prevents bird migration.
D. Predatory animals provide a degree of population control on migratory birds.

Question 84

Tim only owns black clothes – except for sports clothes, where he also owns some green ones. His parents then gave him a pair of blue trousers for his birthday.

Select true or false for each of the following statements

A. Tim has no coloured clothes.
B. Tim owns red clothes.
C. Tim only owns suits.
D. Most of Tim's sports clothes are green.

Question 85

A rowing club celebrates their 100th birthday by holding a 2000m race. Anthony, Peter, Eugene, David and John compete against each other.

Anthony finishes the race after David.
Peter is 3 seconds faster than Eugene.
John is one second slower than Eugene, but faster than David.

		John		

Select true or false for each of the following statements

A. David is last to finish.
B. Peter has the fastest time.
C. John is between Eugene and David.
D. Anthony is slower than Peter.

Question 86

In order to ensure a fair distribution of health care funds, healthy behavior must be enforced through the penalizing of unhealthy actions such as smoking or eating unhealthy foods.

Select the strongest argument against the statement above.

A. Penalties rarely work to change behaviour.
B. Health care funds cannot be allocated fairly.
C. Penalizing unhealthy actions restricts freedom of choice.
D. In order to distribute heath care funds fairly, the richest in society must pay more.

Question 87

Scientists investigate the feeding behavior of bears. Bears are omnivores that feed on meat as well as plants. They find that bears will always prefer berries over any other type of food. Scientists believe that this is due to their sweet taste. They also find that bears do not tend to hunt, but rather tend to eat the meat of freshly deceased animals. They have been seen chasing away other predators in order to steal their catch. This suggests that bears actually are more gatherers than hunters and will only actively attack if cornered or if considering their offspring to be in danger.

Which of the following conclusions can be drawn from the above text?

A. Bears frequently hunt small animals like rabbits and squirrels.
B. Bears prefer seeds over berries.
C. Bears tend to abandon their young if threatened.
D. None of the above.

Question 88

Ants are investigated for their social behaviour. Scientists come to the following conclusions:

Some ants are exclusively responsible for construction work.
Some ants take part in hunting and a proportion of those also play a role in protection of the nest.
Male ants exclusively exist for reproduction.

Which of the following most accurately depicts the relationship between the different roles of ants?

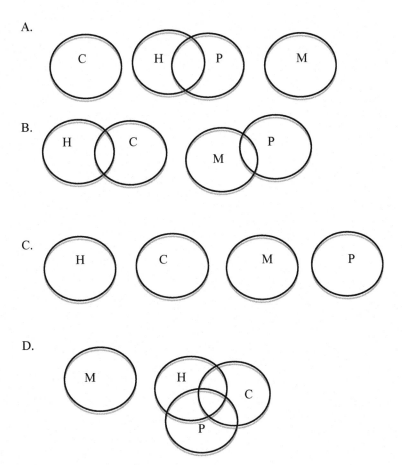

Question 89

Annette is planting a garden. She really likes raspberry jam, so exclusively plants raspberry bushes. There are also some blackberry bushes there. Her husband then decides to plant some cherry trees as well.

Select whether the statements below are true or false.

A. There are no trees in the garden.
B. There is only fruit in the garden.
C. Annette does plant cherry trees.
D. Annette did not plant anything but raspberry bushes.

Question 90

In order to improve our health, we must make an effort to avoid unhealthy foods.

Which of the following arguments most strongly opposes this statement?

A. Health is always dependent on the individual.
B. The definition of unhealthy foods is too imprecise to be of any use
C. This statement encourages fat shaming.
D. All of the above.

Question 91

Stanley decides to build a wardrobe for his new bedroom. He measures the room and finds that he will need to make his wardrobe 75cm wide in order to have space for all his clothes and he decides to make it 2 meters high in order to take full advantage of the ceiling space available. He decides to use the more expensive Walnut wood because he prefers the darker color in comparison to pine. He also decides to use nails instead of screws because he does not have a drill and does not want to buy one. This increases the risk of the wardrobe collapsing.

Which of the following statements is correct?

A. The wardrobe will be large.
B. Nails hold less securely than screws.
C. Pine is darker than Walnut.
D. Peter will need to help him.

Question 92

In order to deliver the highest quality product at the most affordable price possible, a company conducts research into different materials to use for their product. They know that their product needs to be deliver a good compromise of cost, tensile strength, weather resistance and head resistance.

Material A delivers good cost effectiveness and weather resistance, but poor tensile strength and little heat resistance.
Material B delivers good tensile strength, weather resistance and hear resistance, but is very expensive.
Material C is a 50/50 mix of Materials A and B resulting in good tensile strength and excellent weather resistance and heat resistance.

Which of the following conclusions is true?

A. Material A is ideal for a good product.
B. Material B must be steel.
C. Material C must be expensive.
D. Materials A and B do not interact to achieve new properties.

Question 93

A magazine ranks 5 different companies according to annual income.

Company A makes $100,000, which is twice as much as company B.
Company C makes 10% less than company A.
Company D is the smallest company and only makes $75,000.
Company E makes the least.

Which of the following is correct?

A. Company A makes the most.
B. Company B makes more than company C.
C. Company D is amongst the larger companies.
D. Company E must be going bankrupt.

Question 94

Cheryl analyzes the relationships of her friends for a school project. She finds that Sarah gets along well with Tina and Marylin, but not at all with Peter and Astrid. Peter gets along with everybody but Sarah. Tina gets along well with Peter and Sarah, but not with Astrid. Astrid gets along well with Peter.

Which of the following best represents the interaction of Cheryl's friends?

A.

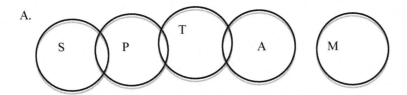

B.

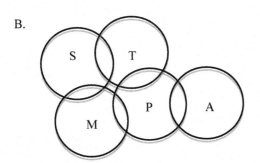

C.

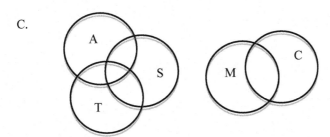

D.

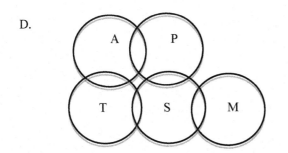

Question 95

A group of scientist investigates the link between population density and the density of newspapers. They find that as populations grow, the amount of newspapers increases until it reaches a ceiling from where it declines as population growth continues. In addition to that they find that the remaining newspapers can increasingly easily be classed into two opposing camps that tend to approach issues from polar opposites.

Which of the following statements is true with regards to the text above?

A. The media is a tool to make us believe in the theories of the politicians.
B. At a certain point population growth is inversely proportional to newspaper density.
C. There are always only two sides to a story.
D. None of the above.

Question 96

A school conducts a survey of their 200 students to find out about their individual job aspirations. They find that they are essentially limited to 5 categories.

10% of children want to be politicians.
40% of children have not chosen a profession yet.
50 children want to be doctors.
20 children want to be actors.
More children want to be teachers than actors.

Which of the following statements is correct?

A. More children want to be politicians than actors.
B. The majority of children want to be doctors.
C. 80 children don't know what they want to be yet.
D. More children want to be politicians than teachers.

Question 97

Philipp and Tom play for Port Town United FC. Philipp scores in every game of the football season. Tom only scores in half the games. In the last game, neither Tom nor Philipp score. In every game Tom scores, Philipp scores at least the same number of goals as Tom. In every game that Tom scores, Port Town never lose. Neither Philipp nor Tom have been sent off in the season.

Which of the following statements is supported by the text?

A. Philipp scores more goals than Tom.
B. Tom scores in the last game of the season.
C. The last game of the season must have been a loss.
D. Tom is better than Philipp.

Quantitative Reasoning

The Basics

The Quantitative Reasoning subtest tests your ability to quickly interpret data and perform relevant calculations upon them. Section 3 contains 36 questions and you have 24 minutes to answer them, giving a total of 40 seconds per question – slightly more generous than the verbal reasoning section. As with all UKCAT sections, you have one additional minute to read the instructions at the start.

There are different types of question you can be asked in Section 3, but all involve interpreting a numerical data source and performing calculations. This is all about testing your natural ability with numbers, how easily you understand numbers and how well you can make calculations based upon new data. You won't find advanced mathematics, so you are at absolutely no disadvantage by not taking A-level maths. Common sources include food menus, timetables, sales figures, surveys, conversion tables and more. All questions have 5 options of which only one is correct.

In this section, the whiteboard you are provided can be useful – use it to scribble down working and intermediate numbers as required.

There is an on screen calculator – a basic calculator for performing arithmetic. You should ideally **practice with a non-scientific calculator** when working through this book, as that will give the closest simulation to what you will get on the day. When you move on to trying the online UKCAT practice papers, the calculator is available on screen as you will see it in the test. In addition to using it to solve questions, practice different calculations to build your speed using it – though this sounds boring, it will save you valuable time on the day. Something many candidates do not realise is that the calculator can be operated using the keyboard controls. Try this out for yourself in practice, and if it works for you it is yet another way to boost your speed when you come to the UKCAT for real. If not then it's fine, it can be used with the mouse too and you will at least be properly prepared, knowing the best approach for you.

Preparation

Be comfortable using the on-screen calculator

As discussed, it's important to know exactly how the calculator works so that you use it quickly and most effectively during the test. Also make sure you know which functions the calculator has, and does not have. Practice using the calculator until you're really familiar with it to ensure you waste no time on the day. Use the automated practice section of the UKCAT website for this – that way, you practice using the same software you will use on the day.

Practice common question styles

Be especially comfortable with things like bus and rail timetables, sales figures, surveys, converting units and working with percentage changes in both directions. These are commonplace in the UKCAT –, but could prove awkward if you're rusty. Likewise be sharp on your simple arithmetic – it might seem basic, but a good knowledge of times tables will save you a lot of time. Even if you're not answering questions, you can hone your skills by practicing reading charts, graphs and tables quickly.

Familiarise yourself with the format of diagrams

Working through plenty of practice questions will help here, as you'll see similar questions coming up again and again. Commonly you will need to use timetables, data tables and different types of graphs to answer Section 3 questions. Make sure you are comfortable with all of these styles of questions.

When looking at an unfamiliar diagram, a clear approach will help you quickly grasp what it shows. Candidates who let the time pressure stop them from properly interpreting the data are much more likely to lose marks. Avoid this common pitfall by following our approach below to quickly read complex data.

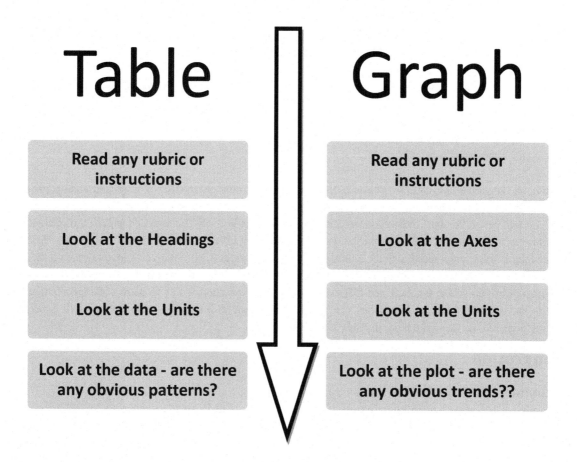

Mental Speed

The main challenge in most questions is finding the right data and selecting the appropriate calculation to perform, rather than the actual calculation itself. However, time is tight so you should be confident with addition, subtraction, multiplication, division, as well as working out percentages, fractions and ratios. Although there is an on-screen calculator in the test, you can save time by doing the basic sums in your head – being confident in your mental arithmetic ability will help you use the time most effectively. There are some good websites and apps to practice quick-fire mental arithmetic – using these you can quickly refresh these essential skills.

Answering Questions

Estimation

Estimation can be very helpful, particularly when the answers are significantly different. If, for instance, answers are an order of magnitude or more away from each other, you can ignore the fine print of the numbers and still get the right answer. If it's a particularly complicated calculation, quickly ask yourself roughly what answer you are expecting. With the simple UKCAT calculator it can be easy to slip up – but if you've already made a quick estimation then you may be alerted to your mistake before you put the wrong answer down. Another important use of estimation is to generate educated guesses if you're short on time. In Section 3 there are five answers per question, so your odds of blindly guessing correctly are low. But here a simple estimation can help. A quick glance or simplified sum might help you eliminate a few answers in only seconds, boosting the chance your guess is correct.

Flagging for review

Flagging for review is so quick and easy, it can always be a useful tool. If you're finding a question difficult, or you've decided it is likely to take too long to solve, put a guess (or quick estimation if possible), flag for review and move on. This allows you to revisit the question at the end if there's time whilst using your time more efficiently elsewhere. When doing this you should make an initial guess, as this ensures you have at least a chance of being correct if you don't have enough time to come back again.

Pace yourself

In this section you have an average of 40 seconds per question, and this is a very useful guide to have. Of course some questions will take more or less time, but you should aim to work steadily forwards at roughly that pace. So after 6 minutes you should be about 9 questions in, and after 12 minutes should have completed about 18 questions and so on. By keeping a regular rhythm to your work, you ensure you don't leave lots of potentially easy questions at the end untouched. It's far better to skip a few tricky questions with a guess to make sure you make a decent effort at all questions, rather than wasting time with the hardest questions and missing out on easier marks.

Read the question first

If the data looks complex, it makes sense to look at the question first before beginning to interpret the data. Just like data-heavy questions in section 1, it can take a few moments to interpret the data provided. By reading the question first, you focus your mind, giving you a better focus to approach the data with and ensuring you only spend time analysing data you actually need to work from.

Top tip! Don't spend too long on any one question. In the time it takes to answer one hard question, you could gain three times the marks by answering three easier questions. *Make the most of every second!*

Example Questions

Example 1

An online company provides personalised sports kit with discounts for bulk purchases. Shipping rates are £4.99. All prices are quoted in pounds and are per item.

No of items	Plain T-shirt	Polo Shirt	Long sleeve T-shirt
1-10	4.99	5.59	5.99
11-50	4.49	5.09	5.49
51+	3.99	4.49	4.99

No of Items	Monotone Print	Multi-tone Print	Embroidered Logo
1-6	0.99	1.99	3.99
7-25	0.49	1.29	3.49
26+	0.29	0.89	2.99

Having read the instructions you know what the table will show. Now look straight at the question below so you know what to do with the data.

A local hockey team requires 26 polo shirts with embroidered logo on the front and printed monotone number on the back. How much will this cost?

A. £194.01
B. £217.62
C. £222.61
D. £225.21
E. £240.81

This is a typical question. Find the right data in the table and start adding it up

Answer: C

This question highlights the need to read the question carefully as to use the correct data from the tables and is a relatively common type of question. If you look closely at the tables, you will realise that the number of items bracketed together changes between the tables. Watch out for this, or similar changes in unit, in the test. If you only skim over the tables you are in danger of missing this and will therefore get the question wrong. However, the calculation required is simple as is often the case, the question is more looking at your ability to pick out relevant data.

Price per polo shirt = 5.09 (base price) + 2.99 (embroidered logo) + 0.29 (monotone print) = £8.37
Price for 26 polo shirts as specified = 8.37 x 26 (number of items required) = £217.62
Total Price = 217.62 + 4.99 (shipping) = £222.61

Example 2

Sarah's route to work consists of an 8 minute walk followed by the 200 bus from Styal centre to Wilmslow station, the train from Wilmslow to Manchester Piccadilly and finally a 6 minute walk to her offices.

200 Bus Timetable				
Manchester Airport	07.05	07.21	07.36	07.49
Styal centre	07.14	07.30	07.45	07.57
Green road	07.18	07.34	07.49	08.01
Wilmslow Leisure centre	07.23	07.39	07.54	08.05
Wilmslow Station	07.31	07.47	08.02	08.13
Wilmslow Centre	07.36	07.52	08.07	08.22
Train Times				
Macclesfield	07.26	07.42	07.58	08.14
Alderley Edge	07.32	07.48	08.04	08.20
Wilmslow	07.40	07.56	08.12	08.28
Stockport	07.54	08.10	08.26	08.42
Manchester Piccadilly	08.14	08.30	08.46	09.02
Manchester Oxford Road	08.23	08.39	08.55	09.11

If Sarah needs to arrive at her offices by 9.00, what time must she leave her house?

A.   07.08
B.   07.24
C.   07.30
D.   07.37
E.   07.49

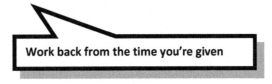

Work back from the time you're given

Answer: D

Bus and train timetables are common questions and it is worth being comfortable working through them logically. On first glance of the question, there appears to be a lot of information. However, if you read the question carefully, you can simply work through the timetables from the end of the question. Split Sally's journey down into stages and consider each in turn – this splits the question so it is easily manageable and makes it less likely to make mistakes.

Sally must be at the offices for 09.00, therefore she must get into Manchester Piccadilly by 08.54 (6 minute walk). The latest train that gets into Manchester Piccadilly by 08.54 arrives in at 08.46. This train leaves Wilmslow Station at 08.12. The latest 200 bus that gets to Wilmslow Station by 08.12 arrives at 08.02. The 08.02 bus to Wilmslow Station leaves Styal Centre at 07.45. It takes Sally 8 minutes to walk to the bus stop therefore she must leave her house at 07.37.

Example 3

The following table shows the rates of income tax per annual income amount. The rates shown apply to James, who earns £10,156 per quarter before tax. Tax is applied to annual income bands as shown:

Annual Income / £	% pay as Income Tax
0 – 10,000	0
10,001 – 25,000	15
25,001 – 41,200	25
41,201 – 50,000	34
50,001 +	40

How much does James earn, to the nearest pound, annually after tax has been deducted?

A. £10,127
B. £30,468
C. £34,468
D. £34,469
E. £40,624

> **Firstly convert James' quarterly income to an annual income to simplify the sums**

Answer: C

Notice the bands of income are not of equal width, and that the question supplies James' income per quarter, but the answer asks for his annual income. Tax is calculated individually per band – no one will pay tax on the first £10,000 earned, the next £15,000 earned is taxed at a rate of 15%, the next £16,200 at 25% and so on.

Firstly you should make life easier for yourself by converting his quarterly income into an annual income. In this question, it is most sensible to work out all calculations as annual income, and annual tax, as this is what the answer requires.

Annual income before tax: 10156 x 4 = £40,624
Tax: income within band x tax rate for that band
Band (0-10000): £0
Band (10001-25000): (25000-10000) x 0.15 = £2250
Band (25001 – 41200): (40624-25000) x 0.25 = £3,906
Total tax: 2250 + 3906 = 6156
Final income after tax deducted: 40624 (total income) – 6156 (tax) = £34,468

Example 4

The following table shows a bank's currency conversion rates between different currencies, with the purchasing currency being listed on the left and the purchased currency being listed across the top. The bank charges a flat fee commission rate of £2.50 on all transactions.

	EUR	AUD	USD	GBP	JPY	INR
EUR	1.00	1.41	1.12	0.75	133	69
USD	0.89	1.26	1.00	0.67	118	62
GBP	1.33	1.88	1.48	1.00	177	92
AUD	0.71	1.00	0.80	0.53	94	49

Peter is travelling to Australia next summer and requires AUD. Using the given currency conversion rates, how much will it cost him in GBP, to the nearest pound, to purchase 650 AUD?

A. £343
B. £346
C. £348
D. £1,220
E. £1,224

> **These answers are in two very different ranges. A quick estimation will narrow down from five to either two or three possible options**

Answer: C

This question is assessing your ability to convert, in this case using currency, but may be using different units of area or such like. Again, it is important to read the question closely to ensure you use the correct currencies, and do not miss the commission rate. Find the row which states the currencies per 1GBP, in this case this is the third row. These are the conversion rates you should be using for your calculation. In this question it is worth simply noting that the amount in GBP will be less than the amount in AUD, as each 1GBP buys you towards 2AUD. This will be useful to check you have converted with the figures in the correct order, as the final answer you get should be less than 650AUD. It is also necessary to look at what currency the commission rate is in. As it is stated in GBP, the easiest way is to convert and then take off the commission rate. This will save you time as it is not necessary to make a further calculation of converting the commission rate to AUD.

When writing your working it is essential to write the currency each value is in as to not get confused. This is the case for any question which requires conversion between different units.

Convert 650 AUD to GBP: 650 ÷1.88 (conversion rate GBP to AUD) = £345.74
Price after commission fee 345.74 + 2.50 (commission fee) = £348.24
The question asks for the answer to the nearest pound, therefore final answer = £348

Quantitative Reasoning Questions

SET 1

The country of Ecunemia has a somewhat complicated tax code. There are four states that make up Ecunemia: Asteria, Bolovia, Casova and Derivia. Each state has its own tax code, including different tax rates on different items. The table below represents the tax a **customer** has to pay when they purchase an item from a store. E.g. a £100 coat in Asteria would cost £110.

	Asteria	Bolovia	Casova	Derivia
Clothes	10%	15%	10%	10%
Food	5%	0%	10%	0%
Imports from other states	20%	5%	10%	15%

The customer must add the tax onto the advertised purchase price. In the case of an item falling into multiple categories (for example, in the case of Imported Food) the higher tax rate is paid and the lower rate is ignored.

Question 1:
A shopper visits a certain supermarket. Without tax, the shopper spends $50 on food, $30 on clothes and nothing on imported items. She spends $88 in total. Which state is this supermarket in?
A. Asteria
B. Bolovia
C. Casova
D. Derivia

Question 2:
Someone runs a supplier in Bolovia, supplying supermarkets in each state in Ecumenia. Each year they supply each state with 250 items of clothing, which the supermarket sells for $40 (including tax), and the supplier gets all of this revenue, minus the tax paid. A competitor in Asteria goes out of business, and this supplier has the opportunity to buy the manufacturing plant for $20,000, and transfer to this state.

If the supplier purchases the site, and moves to Asteria, how many years will it take to make back the cost of purchasing the site?
A. 5 years   B. 12 years   C. 23 years   D. 26 years

Question 3:
John goes into a store and spends $100. Of this, $12 is tax. Which of the following is possible?
A. He shopped in Asteria and bought no imported goods.
B. He shopped in Casova.
C. He shopped in Derivia and bought at least $50 of food (excluding tax).
D. He shopped in Bolovia and spent $10 on imported goods (excluding tax).

Question 4:
Sibella is on a road trip through Ecunemia, driving through different states. On the journey she buys $100 of the finest Asterian ham, $30 of the finest Bolovian caviar, a $10 case of Casovan orange juice and spends $100 on a Derivian dress (all of these prices without tax). Which of the following cannot have been the total amount Sibella spent, including tax?
A. $256
B. $264
C. $273
D. $288

SET 2

As a probe drops through the ocean, the pressure it experiences increases. For every 10 metres the probe drops down, the pressure it experiences increases by 10,000 Pascals (Pa).

Question 5:

A particular probe can survive 200 pounds per square inch without incurring damage. Given that the conversion factor between these units is 7000 Pa = 1 pound per square inch and assuming that pressure at sea level is 0 Pascals, how deep can the probe drop into the ocean without incurring damage?

A. 14 m
B. 140 m
C. 1.4 km
D. 14 km

Question 6:

A different probe is dropped into the ocean and falls downward. This probe can withstand 300,000 Pa of pressure without breaking. A model of the effect of the fluid states that the object's depth in the fluid is $d = \frac{1}{2}\sqrt{(t^3)}$, where d is depth in metres and t is time in seconds. How long will it take for this probe to break?

A. 65 seconds
B. 71 seconds
C. 75 seconds
D. 78 seconds.

SET 3

The fictional drug Cordrazine is used to treat four separate conditions. The following table gives the amount of drug used in each case to treat each condition, written in the form x mg/kg: i.e. for every kilogram you weigh, you take x mg of the drug. The recommended course for the drug is also listed, in the form of number of times a day and how many weeks you need to take the drug.

Condition	Dosage	Course
Black Trump Virus	4 mg/kg	3 times daily for 4 weeks
Swamp Fever	3 mg/kg	Once daily, 1 week
Yellow Tick	1 mg/kg	2 times daily for 12 weeks
Red Rage	5 mg/kg	2 times daily, 3 weeks

Question 7:

Over the course of treatment, John, an 80 kg male, takes 26.88 grams of the drug. Which disease was he prescribed the drug for?

A. Black Trump Virus

B. Swamp Fever

C. Yellow Tick

D. Red Rage

Question 8:

Carol is a 60 kg female who is prescribed the drug (precisely and at different times) three times in one year. Two of the cases are for Yellow Tick. In total she takes 40.32 grams of the drug. Which was the third disease she was prescribed the drug for?

A. Black Trump Virus

B. Swamp Fever

C. Yellow Tick

D. Red Rage

Question 9:

Clarence takes the drug twice in his life. Once he takes it for Swamp Fever at age 18, when he weighs 80 kg, and he takes it later in life at age 40 for Black Trump Virus, when he weighs 110 kg. What is the ratio of the amount he takes each time?

A. 1:23

B. 1:22

C. 1:21

D. 1:20

Question 10:

Danny has liver disease. His system cannot cope with more than 15.5 grams of Cordrazine every 4 weeks. Danny has a medical condition usually treated with Cordrazine, but doctors have advised him to not complete a course of the treatment, as he would exceed the dose that his system is able to cope with. Which of the following statements is possible?

A. Danny suffers from Red Rage and weighs 75 kg.

B. Danny suffers from Swamp Fever and weighs 100 kg.

C. Danny suffers from Black Trump and weighs 45 kg.

D. Danny suffers from Yellow Tick and weighs 75 kg.

Question 11:

Eileen has kidney failure. Her system cannot cope with more than 10 grams of Cordrazine every 4 weeks. She suffers from Red Rage, but doctors have recommended she does not use Cordrazine to treat it, as this would exceed the 10 g dosage her system can cope with. Which of the following weights is the minimum that would support this recommendation?

A. 40.34 kg

B. 42.53 kg

C. 45.81 kg

D. 47.62 kg

SET 4

A bakery sells four varieties of cakes. The cakes contain the following ingredients:

	Sponge (520g)	Madeira (825g)	Pound (710g)	Chocolate (885g)
Flour (g)	125	250	150	200
Butter (g)	125	175	185	175
Egg (g)	120	180	180	120
Milk (g)	25	45	45	150
Sugar (g)	125	175	150	200
Cocoa (g)	-	-	-	40

Question 12:
Which cake contains the highest proportion of flour?
A. Sponge
B. Madeira

C. Pound
D. Chocolate.

Question 13:
The cake recipes are scaled up for a large order. One cake weighs 2.6 kg and contains 625 g of flour. What variety of cake is it?
A. Sponge
B. Madeira

C. Pound
D. Chocolate

Question 14:
Eliza is having a wedding and wants to produce a 4-tiered wedding cake. She wishes each tier to be of different size, and scaled such that that the bottom cake is 50% heavier than normal (e.g. the cake contains 50% more ingredients), the second cake is 25% heavier than normal, the third cake is 10% heavier than normal and the top cake is normal-sized, where each cake is of the same type.

Which of the following is a possible weight of sugar for the cake (rounded to 2 s.f.)?
A. 940 g
B. 970 g

C. 1,000 g
D. 1,030 g

Question 15:
It is known that flour costs £0.55 per 1.5 kg and sugar costs £0.70 per 1 kg. Which of the following is the closest to the cost ratio of flour to sugar in a Madeira cake?
A. 1:2
B. 3:4

C. 4:5
D. 5:6

Question 16:
Milk costs £0.44 per kilogram and flour costs £0.55 per 1.5 kg. What is the cost ratio of flour to milk in a chocolate cake?

A. 1:1
B. 2:3

C. 8:7
D. 10:9

SET 5

The Kryptos Virus is particularly virulent. The infection rate is dependent upon the gender of the recipient. A random sample of 100 men and 100 women are taken from a population and tested for the Kryptos virus using Test A. The results of Test A are displayed below:

	Men	Women
Have virus	45	63
Do not have virus	55	37

Question 17:

What percentage of people tested have the virus?

A. 45%

B. 54%

C. 55%

D. 63%

Question 18:

A population of 231,768 is divided: 53% women, 47% men. Use the data in the table to estimate the number of people in the population that have the Kryptos virus. Assume that the infection rates in each gender will be the same as for the sample population in Test A. Which of the following is the number of people expected to be infected with Kryptos virus in this population?

A. 123,587

B. 123,589

C. 125,541

D. 126,406

Question: 19

3/9 of the men and 5/7 of the women testing positive for Kryptos in Test A have visited the city of Atlantis. Which of the following is the correct percentage of people in the test group testing positive for Kryptos who have **NOT** visited Atlantis?

A. 40%

B. 44%

C. 50%

D. 55%

Question 20:

It is known that Test A is not always correct. Test B is a more accurate test. The 45 men who tested positive for the Kryptos virus using Test A were then re-tested with Test B - only 20 tested positive. Assuming the same proportion of men and women experienced false positive results with Test A, how many women in the test group do we expect to actually have the Kryptos virus?

A. 20

B. 28

C. 35

D. 42

Question 21:

It is decided the women who tested positive under test A should be retested using test B. This time 29 women test positive for the Kryptos Virus. Considering both the men and women tested, what percentage of people who tested positive in Test A also tested positive in Test B (to the nearest whole number)?

A. 40%

B. 45%

C. 50%

D. 55%

SET 6

A business has 3 manufacturing plants and 3 stores. Each plant can ship to each store, and the following table shows the flat rate cost, in pounds sterling (£), of the business sending a truck from the plant to the store.

	Store 1	Store 2	Store 3
Plant A	100	190	530
Plant B	120	180	600
Plant C	140	200	450

Question 22:

Currently the businesses strategy is to send material from Plant A to store 2, from Plant B to store 3 and from Plant C to store 1. One truck is sufficient for a day's delivery. What is the daily cost of this plan?

A. £850
B. £930
C. £970
D. £1,030

Question 23:

The store wishes to optimize their shipping costs by sending material from Plant C to store 3, noticing that the delivery cost is lower. They then choose the two other options that save the most money. What percentage saving is achieved by this strategy relative to the strategy in the previous question (to the nearest whole number)?

A. 18%
B. 20%
C. 22%
D. 24%

SET 7

The table below shows the number of books sold by a bookshop in one day:

	Below 18	Above 18
Non-Fiction	12	30
Horror	50	45
Sci-Fi/Fantasy	23	90
Other Fiction	103	159

Question 24:

The shop also ran an author's visit event in the evening in which 106 people purchased the author's book. These books are **NOT** counted in the above table. What proportion of the books sold on this particular day were sold at the author's visit event (to the nearest whole number)?

A. 13%
B. 17%
C. 21%
D. 25%

Question 25:

Non-fiction books cost, on average, £10, and fiction books cost, on average, £6. What percentage of the shop's revenue (excluding the author's visit event) came from non-fiction books?

A. 10%
B. 13%
C. 19%
D. 23%

Question 26:

Assume that the shop makes this number of sales of each type of book every day. One week, the shop adopts a new marketing strategy and markets non-fiction books more heavily. The result is that the number of non-fiction sales double during this week, but all of the other book sales stay in line with previous sales. How much does the shop earn this week?

A. £24,250
B. £25,620
C. £26,950
D. £27,890

Question 27:

The following week, the shop decides to market the horror books more heavily, resulting in the sales of horror books doubling, and the sales of non-fiction books returning to the normal level. How much does the shop's income increase this week compared to the non-fiction marketing week? Sales of all other books can be assumed to be the same as un-marketed weeks.

A. 1%
B. 2%
C. 3%
D. 4%

SET 8

The following table shows the taxing structure for Italian city hotels:

City	Tax
Venice	1 euro per star per room per night. Rooms with children under 16 are tax exempt.
Rome	Per person, per night: 5 euros for 3 star, 6 euros for 4 star, 7 euros for 5 star, up to a maximum of ten nights worth, after which no tax is charged. Rooms with children under age 10 are tax exempt.
Padua	Per person, per night: 2 euros for 3 star or below, 3 euros for 4 star or above. Rooms with children under 16 are tax exempt.
Siena	2 euros per person per night in high season, 1 euro per person per night in low season. Rooms with children under 12 are tax exempt.

Unless specifically mentioned, assume that all of the people below are aged 18 or over.

Question 28:
A family goes on a tour of Italy in the High season. They are 2 adults and 2 children, aged 9 and 13. They spend two nights in each of Venice, Rome, Padua and Siena. They stay in 3 star hotels for the entire trip, and have two rooms (an adult room and a child room). How much tax do they pay for their trip?
A. EUR 35
B. EUR 56
C. EUR 60
D. EUR 65

Question 29:
Claude is comparing cities. He can either spend 7 nights in Rome in a 4 star hotel, or 8 nights in Padua in a 5 star hotel. Which of the following is the ratio between the tax he pays in Rome and the tax he pays in Padua?
A. 8:3
B. 7:4
C. 6:2
D. 1:4

Question 30:
Alice goes on a trip for 2 days to Venice in a 3 star hotel and for 3 days to Padua in a 4 star hotel. What is the percentage more tax she pays in Padua relative to Venice?
A. 25%
B. 50%
C. 75%
D. 100%

Question: 31
How long does Reuben have to stay in a 4 star hotel in Rome so that the tax would be less than or equal to the tax he incurs if staying the same length of time in a 4 star hotel in Padua?
A. 10 days
B. 15 days
C. 20 days
D. 25 days

SET 9

Peter is building a house that contains rooms of different sizes. The sitting room is 10m x 20m, the hallway is 3m x 10m, and the master bedroom is 15m x 15m. In addition, the house has another square-shaped bedroom, a kitchen and a bathroom.

Question 32:

Assuming that the second bedroom walls are 60% of the length of the master bedroom, what is the area of the second bedroom?

A. 64 m^2
B. 81 m^2
C. 100 m^2
D. 121 m^2

Question 33:

Suppose the kitchen has a floor area of 100 m^2 and the bathroom has a floor area of 4 m^2, and the second bedroom has the floor area calculated in the previous question. What percentage of the area of the house is taken up by the master bedroom?

A.   30%
B.   35%
C.   40%
D.   45%

Question 34:

After building the house, Peter decides to add an extension to the sitting room, turning it into a combined lounge and dining room. He extends the room by increasing the length of the longer wall by 5 metres. The lounge is 3 metres high. How much extra wall (in m^2) does Peter have to build, assuming that he is extending directly outwards and cannot move or re-use any wall?

A. 15m^2
B. 30m^2
C. 45m^2
D. 60m^2

Question 35:

A larger extension is considered, and two builders offer Peter separate quotes. The first builder offers to build wall at a cost of £15 per m^2, but there is also a flat fee of £200 just for starting the job. The second builder offers to build wall at a cost of £16 per m^2 but with no flat fee at the start. If Peter builds 300 m^2 of wall, what is the ratio of builder 1 cost to builder 2 cost (to 3 s.f.)

A.   1.00:1.00
B.   1.00:1.02
C.   1.02:1.00
D.   2.00:3.00

SET 10

The table below shows the service prices for competing mobile phone plans A-D. Any SMSs or call minutes beyond those free with the plan are charged individually at listed price.

	A	B	C	D
Monthly fee	£0	£5	£10	£15
# Free SMSs	0	200	1000	Unlimited
# Free call minutes	0	0	100	Unlimited
Price/SMS	10p	20p	20p	-
Price/call minute	10p	20p	20p	-

Question 36:

John buys Plan B for one month and calls for 15 minutes and sends 207 SMSs. How much does he pay this month?

A. £5.00

B. £6.60

C. £7.80

D. £9.40

E. £11.20

Question 37:

Robin buys Plan A, and makes no calls. How many SMSs can Robin send before Plan B would have been cheaper?

A. 6

B. 21

C. 51

D. 101

E. 121

Question 38:

Mary wants to call for 5 minutes and send 5 SMSs every day in September. Which plan should she choose for the lowest cost?

A. A

B. B

C. C

D. D

E. A and B are both the lowest

Question 39:

Evan buys Plan B but Chris buys Plan C. Which of these options is cheaper for Evan than Chris per month?

A. They each call for 29 minutes and send no SMSs.

B. They each call for 26 minutes and send 174 SMSs

C. They each send 351 SMSs and call for 8 minutes.

D. They each call for 4 minutes every day of the month.

E. They each send 223 SMSs and make no calls.

Question 40:

Rachel doesn't send any SMSs and buys Plan C. What is the maximum percentage by which she can exceed her free call minutes allowance without Plan D being cheaper?

A. 5 %

B. 10 %

C. 25 %

D. 50 %

E. 75 %

SET 11

A muffin recipe calls for ingredients in the amounts listed in the table below:

Ingredient	Density	Amount
Flour	600 gram/dm^3	2 cups
Sugar	850 gram/dm^3	1 cup
Milk	1050 gram/dm^3	½ cup
Butter	950 gram/dm^3	4 tablespoons

1 cup = 2.5 decilitres (dl); 1 tablespoon = 15 millilitres (ml); 1 cubic decimetre (dm^3) = 1 litre

Question 41:
How many cups of ingredients are called for overall by the recipe (to 2 decimal places)?
A. 3.54
B. 3.66
C. 3.74
D. 3.82
E. 3.86

Question 42:
What weight ratio of milk to butter does the recipe call for (to 1 decimal place)?
A. 2.3:1
B. 2.7:1
C. 3.1:1
D. 3.4:1
E. 3.9:1

Question 43:
Jane wants to use only a ½ cup measure for baking. What is the smallest number of cups of flour she would need for it to be possible to measure all required ingredients in ½ cups?
A. 2
B. 10
C. 25
D. 30
E. 50

Question 44:
To make pancakes, the amount of flour and milk are reversed. What is the average density of pancake batter, assuming that there are no interactions that change the densities of the individual ingredients when they are mixed?
A. 930 grams/dm^3
B. 970 grams/dm^3
C. 1,050 grams/dm^3
D. 1,070 grams/dm^3
E. 1,100 grams/dm^3

Question 45:
If Peter wanted to make 10 muffins weighing 100 grams each, how much butter would he need to 1 decimal place? Assume that the finished product weighs the same as the initial dough.
A. 55.1 grams
B. 62.3 grams
E. 81.3 grams
C. 70.7 grams
D. 76.4 grams

Question 46:
When Peter's ten 100 gram muffins are done, assuming no losses to cooking, what percentage of the weight will be made up by flour, to the nearest whole number?
A. 35 %
B. 39 %
C. 43 %
D. 46 %
E. 52 %

SET 12

New ocean crust is formed at spreading ridges. The area of the crust formed is dependent on temperature. The volume of crust formed in a given time interval depends on the **crust cross sectional area** and on the spreading rate (the rate at which newly formed crust moves away from the spreading ridge, an independent variable).

The relationship between **crust volume** formed in a time interval, **cross sectional area** and **spreading rate** is:

Crust volume per time = cross sectional area x spreading rate

The table below gives the crustal cross sectional area, spreading rate and temperature at Locations A-D:

	A	B	C	D
Cross Sectional Area (km^2)	10	20	30	40
Spreading rate (mm/year)	150	20	100	50
Temperature (°C)	1300	1400	1500	1600

Question 47:
Assuming that the trends in this table can be reliably extrapolated, at which temperature would the crust volume formed in a year be expected to be 0 km^3?

A. 1,200 °C
B. 1,400 °C
C. 1,600 °C

D. 1.800 °C
E. 2,000 °C

Question 48:
If the temperature at Location A increased by 50%, what would be the spreading rate?

A. 25 mm/year
B. 50 mm/year
C. 100 mm/year

D. 150 mm/year
E. 225 mm/year

Question 49:
What volume of crust is formed in a year at Location B?

A. 400 m^3
B. 400 °C km^3
C. 40,000 km^3

D. 400,000 m^3
E. 560,000 °C km^3

Question 50:
If the spreading rates of Locations A and C were exchanged, what would be the ratio of crust volume formed at the two locations each year (to 1 decimal place)?

A. 1:1.0
B. 1:3.3
C. 1:4.5

D. 1:5.6
E. 1:6.0

Question 51:
If the same crustal volume was produced in the same amount of time at 2 locations, E with temperature 1300 °C and F with temperature 1450 °C, how many percent faster/slower was the spreading rate at location E than F?

A. 250 % faster
B. 25 % slower
C. 400 % faster

D. 40 % slower
E. 500 % faster

Question 52:
If the temperature at Location D was decreased by 10%, what would be the crustal volume formed in 3 years?

A. 2,000 m^3
B. 2,000 °C km^3
C. 3,200 km^3

D. 3, 200,000 m^3
E. 3, 600,000,000,000,000 mm^3

SET 13

A new drug to treat vision problems in diabetics is tested on volunteers. It is also tested on control groups of diabetics without vision problems and healthy volunteers with or without vision problems. Some volunteers are given one inactive placebo pill which they are told is the drug. There are the same number of people in each group testing either the drug or placebo, as indicated below.

The table below shows the number of volunteers in Groups A-D who self-reported improved vision and their measured average accuracy reading letters before and after taking the drug or a placebo.

	Group A		Group B		Group C		Group D	
	Drug	Placebo	Drug	Placebo	Drug	Placebo	Drug	Placebo
Number Improved	15	9	8	6	9	7	7	8
Accuracy Before (%)	27	27	60	60	29	29	68	68
Accuracy After (%)	36	31	66	61	31	32	70	70

Group A: 50 diabetics with vision problems (25 in each group)
Group B: 46 diabetics without vision problems (23 in each group)
Group C: 44 healthy volunteers with vision problems (22 in each group)
Group D: 48 healthy volunteers without vision problems (24 in each group)

Question 53:
What is the average percentage of participants who self-report vision improvements after receiving an inactive pill to the nearest percent?
A. 26 %
B. 31 %
C. 32 %
D. 33 %
E. 36 %

Question 54:
By what ratio is visual accuracy in reading letters increased by the drug in diabetics with poor sight relative to healthy volunteers with poor sight (to 2 decimal places)?
A. 1:0.78
B. 3.50:1
C. 4.21:1
D. 4.50:1
E. 4.83:1

Question 55:
If there are 10 women in Group A and their average accuracy was 45 % after receiving the drug, what was the average accuracy of the men in the group after receiving the drug?

A. 16 %
B. 27 %
C. 30 %
D. 36 %
E. 41 %

	Group A		Group B		Group C		Group D	
	Drug	Placebo	Drug	Placebo	Drug	Placebo	Drug	Placebo
Number Improved	15	9	8	6	9	7	7	8
Accuracy Before (%)	27	27	60	60	29	29	68	68
Accuracy After (%)	36	31	66	61	31	32	70	70

Group A: 50 diabetics with vision problems (25 in each group)
Group B: 46 diabetics without vision problems (23 in each group)
Group C: 44 healthy volunteers with vision problems (22 in each group)
Group D: 48 healthy volunteers without vision problems (24 in each group)

Question 56:
If the general population has 100 000 diabetics with vision problems, how many of these people would be expected to self-report improvements in their vision because of the effects of the drug?

A. 24,000 people
B. 32,000 people
C. 36,000 people
D. 60,000 people
E. 96,000 people

Question 57:
When the drug dose was doubled, the placebo groups showed no change in numbers or accuracy, but the number of Group A volunteers who reported improved vision jumped to 18. Assuming that drug effectiveness is dose dependent, what percent of volunteers in Group A taking the drug would be expected to self-report improved vision if the dose was tripled?

A.   54.0 %
B.   72.0 %
C.   84.0 %
D.   90.0 %
E.   100.0 %

Question 58:
Which of the following statements is supported by the data in the table?

A. The placebo is more effective than the drug.
B. The drug acts to improve vision in diabetics and healthy volunteers.
C. Volunteers who see well are more motivated to improve vision than those with vision problems.
D. Thinking you have taken a drug to improve vision improves your vision.
E. The data are inconclusive.

SET 14

Dave weighs 200 pounds and has a Basal Metabolic Rate (BMR) of 2000 calories. Elizabeth weighs 140 pounds and has a BMR of 1500 calories. The table below shows the calorific value of the foods they eat:

	Cereal	Sandwich	Apple	Chocolate	Lasagna	Chicken	Vegetables
Calories	400	500	100	350	700	250	200

To lose one pound of fat requires a 3500 calorie deficit, obtained by eating fewer calories than the BMR or burning calories by exercising. Running burns 5 calories per hour per pound you weigh at any running speed. Cycling burns calories according to the following relationship, where M is mph cycling speed:

Calories burned per mile = 50 calories + (5 calories x (M-10))

Question 59:

Dave wants his workout to take one hour on a 5 mile track. What is the maximum number of calories he can burn by running or cycling?

A. Burn 125 calories running
B. Burn 125 calories cycling
C. Burn 1,000 calories running

D. Burn 1,000 calories cycling
E. Burn 1,250 calories running

Question 60:

Dave doesn't want to eat less than his BMR and can only run for 30 minutes a day, but cycles 20 miles every day in an hour. How long will it take him to lose 10 pounds?

A. 5 days
B. 7 days
C. 10 days

D. 14 days
E. 30 days

Question 61:

Elizabeth and Dave both want to lose 10% of their body weight without dieting or cycling. What is the ratio of minutes a day Elizabeth would have to run to those Dave would have to run to achieve their goal at the same time to 1 decimal place?

A. 1:0.5
B. 1:0.7
C. 1:1.0

D. 1:1.4
E. 1:2.0

Question 62:

If Elizabeth eats cereal for breakfast, a sandwich for lunch, chicken and vegetables for dinner and does no exercise, in how many full days will she have reached her goal of 10% weight-loss?

A. 327 days
B. 354 days
C. 372 days

D. 416 days
E. 435 days

Question 63:

If Elizabeth also began cycling 10 miles in 1 hour every day, how much faster would she reach her goal than in question 62?

A. 1.00
B. 2.50
C. 3.00

D. 3.33
E. 4.33

Question 64:

Elizabeth eats one chocolate everyday; 3 times as much chicken as chocolate and twice as much cereal as chicken. If she exchanged these foods with 3 different foods in the table in the same proportions, what is the ratio of her rate of weight change before and after the switch, assuming she is trying to obtain the lowest weight she can?

A. 1:1   B. 1:2   C. 1:5   D. 2:1   E. 5:1

SET 15

Visitors to an amusement park pay for food, rides and games with coupons. Coupons can be bought individually for £1 each or in multipacks at a discounted price. A £70 wristband can gives free entry and access to all rides (but not games or food) without using coupons. The table below shows the cost in coupons for each activity:

	Entrance	Rollercoaster	Fun House	Swings	Carnival Games	Candy Floss
All Day	10	4	2	3	1	2
Night	5	3	2	2	1	2

Question 65:

Susan buys a 20 coupon multi-pack at 10 % off single coupon price. She rides the rollercoaster five times at night. What percent off did she get on the first rollercoaster ride compared to buying single day tickets (to 1 decimal place)?

A. 10.0 %

B. 22.5 %

C. 25.0 %

D. 32.5 %

E. 33.3 %

Question 66:

One weekend the single coupon prices are raised 20 %. Greg wants to ride the rollercoaster 10 times, buy 3 candy floss, play a carnival game, go through the fun house 2 times and ride the swings during the day. What is the ratio of the cost with a wristband to the cost without a wristband to 2 decimal places?

A. 1:0.93

B. 1:0.98

C. 1:1.02

D. 1:1.10

E. 1:1.12

Question 67:

Andy went to the amusement park one night. He rode the rollercoaster 50 % more times than he rode the swings, and rode the swings 20 % more times than he played carnival games. He used a whole number of coupons that were cheaper than getting a wristband. How much did Andy pay?

A. £31

B. £44

C. £49

D. £56

E. £70

Question 68:

Anna and James each spent one pound less than the cost of a wristband on single coupons one day at the amusement park. Anna went on the rollercoaster for half of her rides and the fun house for the rest. James went on the swings every odd ride and in the fun house every even ride. Neither of them went on any other rides or bought any food. What is the ratio of the number of rides Anna went on to the number James went on?

A. 1:0.78

B. 1:0.81

C. 1:1.23

D. 1:1.28

E. 1:2.56

Question 69:

A 10-weekend season pass covers all costs in the park and is available for £1,000. Erik goes to the park one day every weekend and buys a wristband each time. On the first weekend he buys one candyfloss, and the next four weekends increases the number of candyfloss he buys by 100%, relative to the previous weekend. The four weekends after that he increases the number of candyfloss he buys by 50% each weekend, relative to the previous weekend. On the 10th weekend he is sick of candyfloss and buys none.

What is the ratio of the cost without and with a season pass (to 2 decimal places)?

A. 0.7 : 1

B. 0.92 : 1

C. 1 : 0.92

D. 1 : 1.15

E. 1.15 : 1

SET 16

The table shows the prices a pizzeria charges for their pizza:

Type	Italian			Pan Pizza		
	Cheese	Toppings	Cheese	Topping	Stuffed Crust
Price Small	£6.00	+50p/topping	£10.00	+50p/topping	+£1.00
Price Medium	£8.00	+£1.00/topping	£12.00	+£1.00/topping	+£1.50
Price Large	£10.00	+£1.50/topping	£14.00	+£2.00/topping	+£2.00

The pizzeria also offers three discount deals: 20% off orders from £20 - £29.99, 30% off orders from £30 - £49.99, and 50% off orders over £50. Small pizzas have 6 slices, medium pizzas have 8 slices, and large pizzas have 10 slices

Question 70:

Josh gets 2 large stuffed-crust 2 topping pan pizzas, 1 medium 3 topping Italian pizza and 3 small cheese pan pizzas. How much does he pay?

A. £31.00

B. £40.50

C. £52.50

D. £81.00

E. £96.00

Question 71:

Janet bought cheese pan pizzas for the cheapest cost per slice and got £35 off as a discount deal. How many slices did she buy?

A. 35

B. 48

C. 50

D. 64

E. 70

Question 72:

Joey bought some plain cheese pizzas for a total price of £60 post-discount. All the plain cheese pizzas were the same. What is the price of the one type of pizza he could **NOT** have bought?

A. £6.00

B. £8.00

C. £10.00

D. £12.00

E. £14.00

Question 73:

Lea always buys 30 slices of cheese pan pizza with 2 toppings and stuffed crust. What is the ratio of the cost of buying all large to all small pizzas (to 2 decimal places)?

A. 1:0.50

B. 1:0.75

C. 1:0.80

D. 1:1.00

E. 1:1.25

Question 74:

Kate and two friends each order cheese pan pizzas (with no toppings) and get 30% off their order. They ordered the pizzas to pay the smallest price which gets this discount, but ended up with 25% more slices than they could eat. How many slices did they manage to eat?

A. 16

B. 18

C. 20

D. 25

E. 32

SET 17

Sarah has three journalist jobs she splits her time across. The table shows a breakdown of what she earns at each job. Her salary is composed of a fixed starter wage she earns for showing up and an hourly wage on top of that. Her hourly wage increases in each job the more hours she works at that job. She must pay for her own travel expenses. Each travel cost occurs once per job she completes, and is not affected by the length of the job in hours.

	Travel Cost	Fixed Starter Wage	Hourly Wage	Average Job Length	Hourly Wage Growth
Job A	£5	£10	£10 per hour	2 hours	£5 per 50 worked
Job B	-	£5	£15 per hour	1 hours	£10 per 100 worked
Job C	£10	£5	£20 per hour	4 hours	£5 per 100 worked

Question 75:

If Sarah works for 1 hour at each job, what will be the ratio of the earnings expressed as [Job A Earnings:Job B Earnings:Job C Earnings] (to 2 decimal places)?

A. 1.00:0.75:1.00

B. 1.00:1.00:0.75

C. 1.00:1.25:1.00

D. 1.00:1.33:0.75

E. 1.00:1.33:1.00

Question 76:

Sarah worked 25 2-hour jobs, 4 1-hour jobs and 1 4-hour job for Job A in her first month. How much did she earn?

A. £625.00

B. £730.00

C. £770.00

D. £980.00

E. £1,020.00

Question 77:

Sarah can work 50 2-hour jobs per month. For which single job should she work these hours to earn the most from 2 hour jobs at the end of the month?

A. A

B. B

C. C

D. A and B are the same

E. B and C are the same

Question 78:

Sarah pays 10% income tax if her monthly salary exceeds £1275. How many hours should she work in her first month for Job C, if all jobs are the average job length, to earn the highest amount possible whilst not paying tax, to the nearest half-hour?

A. 60.0 hours

B. 62.5 hours

C. 65.0 hours

D. 68.0 hours

E. 75.0 hours

Question 79:

At the start of her third month, Sarah has worked 200 hours at Job C. She works 100 hours at average job length this month. How much of her month's earnings go to 10% income tax?

A. £150.00   B. £187.50   C. £200.00   D. £287.50   E. £300.00

Question 80:

Job B wants her to work a minimum of 50 hours a month for them, and Job A and Job C require that she works at least the same hours for them as she does at any other jobs she has, or no hours at all. Assuming all jobs are the average job length, which arrangement would give her maximum earnings in her first 100-hour work month?

A. 50 hours for A and 50 hours for B

B. 50 hours for A and 50 hours for B

C. 50 hours for B and 50 hours for C

D. 100 hours for B

E. None of the above.

SET 18

The table below shows the number of cars passing a toll booth going into the town centre and how many passengers the cars carried, including the drivers. It also shows the number of passengers who got off at the town's central underground station each day.

	Mon	Tue	Wed	Thu	Fri	Sat
Number of cars	1,517	1,632	987	1,465	2,024	478
Total car passengers (incl. Drivers)	1,873	2,421	1,116	2,101	2,822	1,339
Underground passengers	2,346	1,798	3,103	2,118	1,397	576

Question 81:

Taking the underground costs £5. On Wednesdays this fare is reduced 15% and it is cheaper for some drivers to leave the car at home. How much more revenue is generated on Wednesday than the next highest grossing day?

A. £1,457.75

B. £3,537.25

C. £5,275.75

D. £1,1730.00

E. £1,3187.75

Question 82:

On weekdays, what is the ratio of the average number of people being driven (not driving) in cars to the average number of people riding on the underground (to 2 decimal places)?

A. 1:0.25

B. 1:0.78

C. 1:3.97

D. 1:7.60

E. 1:9.60

Question 83:

What is the ratio of the average number of people per car on Tuesday compared to the average number of people per car on Saturday?

A. 1:0.67

B. 1:0.85

C. 1:1.27

D. 1:1.33

E. 1:1.89

Question 84:

What is the ratio of the number of Underground passengers on Monday compared to that on Saturday?

A. 1:0.13

B. 1:0.25

C. 1:0.40

D. 1:1.50

E. 1:4.08

Question 85:

The tollbooths charge £4 per car and an additional £1 per passenger (including the driver). 80% of this payment is tax. How much tax is paid at the tollbooths next week from Monday-Friday if there are 4,219 commuters everyday split in the ratio of 2:1.7:1 - underground passenger : car passenger (including driver) : car ratio?

A. £1,938.45

B. £4,219.00

C. £4,560.50

D. £5,198.40

E. £6,498.96

SET 19

Music practice rooms are available seven days a week, with each day being split into three sessions: morning (8am-2pm), day (2pm-8pm) and night (8pm-2am). The table below gives the prices for the hourly rental of the music practice rooms. Some information is missing. The "two sessions" column indicates the hourly charge if two sessions are booked on the same day.

Type of room	Deposit	Cost per Hour			
		1-2 hours (night session)	3-6 hours (night session)	Two sessions	All day
Basic (no piano)	£10.00	-	-	£11.00	£8.00
Standard (upright piano)	£25.00	£20.00	£18.00	£16.00	£12.00
Superior (baby grand piano)	£50.00	£30.00	£26.00	£22.00	£16.50
Deluxe (grand piano)	-	£45.00	£38.00	£30.00	£20.00

NB: All prices above include VAT (25%)

Question 86:
The hourly rate for a day session is 10% more expensive than a night session. What is the total cost, excluding the deposit, for 3 hours in the Superior room during a day session?
A. £59.40
B. £70.20
C. £78.00
D. £84.18
E. £85.80

Question 87:
The total cost for two 6 hour sessions in the Deluxe room is £460. How much is the deposit?
A. £64
B. £75
C. £82
D. £100
E. £136

Question 88:
Mike books a Basic room and a Standard room for a full night session. The total cost is £221 including deposit. What is the hourly rate for a full night session in a Basic room?
A. £12.50
B. £13.00
C. £15.00
D. £16.83
E. £18.83

Question 89:
The hourly rate for a morning session up to 6hrs is 5% cheaper than a night session. The deposit remains the same. A Superior room is booked for 90 minutes one morning, all costs paid up front. How much is paid at the start of the session? (Assume half hours can be booked at half the hourly rate.)
A. £97.25
B. £92.75
C. £95.00
D. £90.50
E. £90.25

Question 90:
A Basic room is booked for 18 hours each day for three full weeks. What is the total cost of this booking excluding VAT and deposit?
A. £1,935.36
B. £2,419.20
C. £2,459.52
D. £3,024.00
E. £3,074.40

SET 20

A group of 180 people took part in a perception study and were asked to count how many differences they could spot between two similar pieces of short video footage. The results are given below

		Age (years)					
		10 to 16	16 to 22	22 to 34	34 to 48	48 to 65	65+
Differences correctly spotted	<5	9	10	10	16	15	19
	5 to 10	7	12	9	8	8	5
	11 to 15	11	8	6	2	8	9
	15+	3	2	0	1	2	0

Question 91:

What percentage of people under the age of 22 spotted more than 10 differences?

A. 31.3%　　　　B. 33.3%　　　　C. 38.7%　　　　D. 46.7%　　　　E. 63.2%

Question 92:

75% of the results for the people who spotted 5 to 10 differences correctly were removed from the study. What percentage of the remaining people aged 16-22 spotted more than 15 differences?

A. 6.3%　　　　B. 6.9%　　　　C. 8.5%　　　　D. 8.7%　　　　E. 9.4%

Question 93:

25% of people who correctly spotted over 10 differences, also *incorrectly* spotted over 10 differences. How many people was this?

A. 11
B. 12
C. 13
D. 14
E. 15

Question 94:

10,000 people aged 48 or older take this test. Using the data, estimate how many spotted fewer than five differences to the nearest 50.

A. 2,300
B. 2,900
C. 4,500
D. 5,100
E. 5,150
F. 5,200

Question 95:

The test is repeated with the same population. The number of 16-34 year olds who spot 11-15 differences increases by 50%. All other age groups experience no change. What is the new ratio between 16-34 year olds and the total number of people in the other age groups who spot 11-15 differences?

A. 1:3
B. 4:17
C. 14:44
D. 14:51
E. 21:51

SET 21

The pie chart below shows the favourite sports of some high school students. Every student plays only their favourite sport in games lessons. The school has 1300 students, with an exact 50:50 split between boys and girls.

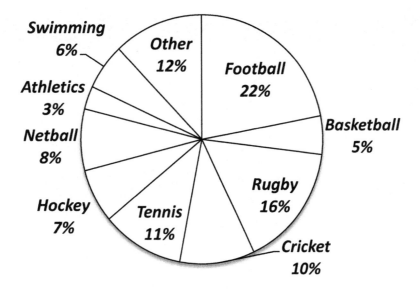

Question 96:

What is the difference between the number of boys that play football and the number that play netball in games lessons?

A. 90	C. 104	E. 182
B. 91	D. 180	F. 286

Question 97:

The senior football teams are picked from the two most senior years – a total of 350 students. Only those whose favourite sport is football play. At least 11 people are needed per team. What is the maximum number of teams that could be made? Assume that the values given in the chart are representative of these years.

A. 4 B. 5 C. 6 D. 7 E. 8

Question 98:

All those whose favourite sport is basketball are boys and all those whose favourite sport is netball are girls. 80% of the basketball boys are invited to play netball. What proportion of the netball-playing population do they then make?

A. 17% B. 25% C. 33% D. 42% E. 50%

Question 99:

One quarter of students in the *Other* category have a favourite sport which is a ball sport. In the whole school, how many students have a favourite sport which is a ball sport?

A. 39 B. 117 C. 572 D. 1,066 E. 1,183

Question 100:

Only boys play cricket. Only girls play hockey. The gender split for tennis follows that of the school as a whole. How many more boys play cricket or tennis than girls play hockey or tennis?

A. 39 B. 58 C. 59 D. 111 E. 112

SET 22

The number of apples picked by a company per year is given below, along with the quality of the apples. 30% of edible apples are sold as they come. Passable apples and the remaining edible apples are processed into cider.

Apples which are No Good are not used for human consumption, and are instead discarded for animal food.

	1998	1999	2000	2001	2002	2003
Edible	1,100,547	1,398,663	1,563,327	1,443,599	1,763,870	1,931,784
Passable	2,983,411	2,691,553	3,008,941	2,790,456	2,651,399	2,439,012
No Good	400,001	391,747	398,014	568,440	494,309	571,221

Question 101:

What is the percentage increase in the number of apples used for human consumption from 1998 to 2003?
A. 7%
B. 10%
C. 22%
D. 76%
E. 93%

Question 102:

What percentage of all No Good apples was produced in the year most apples could not be used for humans?
A. 11.6%
B. 11.8%
C. 13.9%
D. 19.9%
E. 20.2%

Question 103:

2004 saw a three-fold increase on 2003 in the number of No Good apples. The total number of apples fit for consumption remained the same. What was the difference in number between processed and No Good apples in 2004 to the nearest apple?
A. 2,077,598
B. 2,224,675
C. 2,478,954
D. 2,675,133
E. 2,765,131

Question 104:

The next six-year period saw an overall 20% increase on the period 1998-2003 in the total number of edible apples picked. How many were sold as they came between 2004 and 2009?
A. 3,588,698
B. 3,321,646
C. 3,312,644
D. 2,392,465
E. 2,208,430

Question 105:

Generally, 20 apples give 1 litre of cider. Given that 2004 saw the same number of apples fit for human consumption as 2003, roughly how many litres of cider were produced in 2004?
A. 122,000 l
B. 189,600 l
C. 215,400 l
D. 247,100 l
E. 988,400 l

SET 23

Jen tracks her daily jogs using an app which gives her data on her performance. Her app tells her that her average speed is 5 mph.

Conversion factor: 1 mile = 1.6 km

Question 106:

On wet days, Jen's average speed decreases by 8%. How many kilometres does she cover in 40 minutes?

A. 3.1 km
B. 3.3 km
C. 4.9 km
D. 5.3 km
E. 7.4 km

Question 107:

Jen begins training for a marathon (26 miles). She starts off by trying to complete a marathon over the space of four equally long jogs. Estimate how long each jog is. Assume dry conditions.

A. 42 minutes
B. 46 minutes
C. 1 hour 18 minutes
D. 1 hour 25 minutes
E. 1 hour 30 minutes

Question 108:

After starting marathon training, her average speed decreases to her old wet speed; her average wet speed remaining 8% slower than this. Estimate, therefore, how long it would take her to cover 12km in the rain.

A. 1 hour 38 minutes
B. 1 hour 46 minutes
C. 2 hours 17 minutes
D. 2 hours 37 minutes
E. 2 hours 50 minutes

Question 109:

After bringing her average speeds back to their original values, Jen starts a new regime. She goes on four jogs, each being 50% further than the last. Her first jog is 4km long. How long does the final jog take in dry conditions?

A. 1 hour 8 minutes
B. 1 hour 41 minutes
C. 1 hour 50 minutes
D. 2 hours 9 minutes
E. 2 hours 42 minutes

Question 110:

Lots of training later, Jen completes the marathon in a time of 3hrs 42mins on a dry day. What is the percentage increase in Jen's dry average speed compared to her original one?

A. 7%
B. 12%
C. 41%
D. 52%
E. 53%

SET 24

The table below gives the prices per person per week for different luxury holiday accommodations with different swimming facilities. Some types of accommodation offer a choice between swimming facilities. Some information is missing. Additional days are charged at 1/7 of the weekly cost.

	Studio	Apartment	Villa	Palazzo
No pool	£50.00	£70.00	£95.00	£155.00
Shared pool	£60.00	£80.00	-	-
Private pool	-	£100.00	-	£325.00
Beachfront	-	-	£220.00	£480.00

Question 111:

Villas are available with a private pool, and currently they are on sale: 20% off the standard price, where the standard price sits halfway between that of an apartment with a private pool and a palazzo with a private pool. How much would this cost for two people for one week?

A. £323   B. £332   C. £340   D. £415   E. £664

Question 112:

A group of twelve rents out a beachfront palazzo for four weeks. A booking fee is required from each member of the group, in this case charged at 10% of the weekly cost per person. What is the total cost of the booking?

A. £23,040
B. £23,161
C. £23,616
D. £25,344
E. £25,614

Question 113:

A couple rents an apartment with a shared pool for 20 days. The total cost is £492.89. How much is the booking fee?

A. £35.55
B. £35.57
C. £35.75
D. £37.55
E. £37.75

Question 114:

A family of four stays at a beachfront villa for two weeks, with no booking fee. Due to a complaint, they are refunded 20% of the standard charge. How much does the family pay?

A. £1,408
B. £1,760
C. £1,920
D. £2,340
E. £2,620

Question 115:

A company hires three palazzi with private pools for a week for the grand total of £19,500. The booking fee is 10% of the total cost. Assuming each palazzo has the same number of people staying in it, how many people are there in each palazzo?

A. 18   B. 20   C. 22   D. 54   E. 60

SET 25

FastFoodCo is a fast food take-away that delivers directly to customers' homes. Delivery rates are £3.00 for orders less than £10, £1.50 for orders from £10 - £15 and free for orders over £15. Below is a selection from their menu (delivery and food prices exclude 20% VAT, which is payable on all orders). VAT is added after delivery and any discounts have been taken into account.

Item	Cost
Green Curry	£3.95
Chicken Curry	£2.95
Noodles	£2.95
Chicken Tikka	£4.95
Vegetarian Curry	£3.95

Question 116:
John orders a green curry, noodles and a vegetarian curry. What is the total price?
A. £16.62
B. £14.82
C. £14.52
D. £12.35

Question 117:
Katy orders 3 noodles, 2 chicken tikkas and a green curry. Her total is:
A. £22.44
B. £25.67
C. £27.24
D. £29.04

Question 118:
John orders 2 noodles and a vegetarian curry. What is his total price?
A. £15.42
B. £14.24
C. £13.62
D. £12.14

Question 119:
Katy orders a green curry, 3 noodles and a vegetarian curry. What is her total?
A. £21.89
B. £20.10
C. £18.52
D. £18.09

Question 120:
A final 'two for the price of one' offer is applied for noodles. John orders 4 noodles, 2 chicken tikkas and a green curry. What is his total?

A. £18.20
B. £21.33
C. £22.51
D. £23.70
E. £24.31

SET 26

The graph below shows the first quarter profits (in GBP) of four suppliers of prescription medicine.

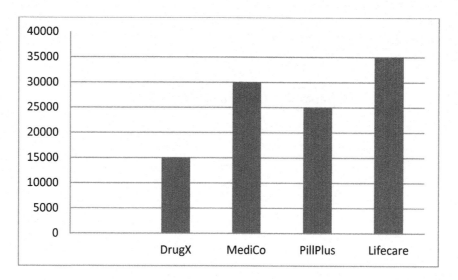

Question 121:
What percentage of total first quarter profits were earned by MediCo?
A. 26.6%   C. 30.4%
B. 28.6%   D. 33.3%

Question 122:
What percentage of the total first quarter profit is from MediCo and Lifecare combined?
A. 56.8%   C. 61.9%
B. 59.5%   D. 62.3%

Question 123:
PillPlus offers a 10% discount on all products in the second quarter. As a result, their sales increase and profit increases by 15%.

Assuming that the profits of all other suppliers remain constant into the second quarter, what percentage of the total second quarter profits did PillPlus make?
A. 16.4%   C. 26.4%
B. 23.8%   D. 27.3%

Question 124:
During the third quarter, all profits fall by 10% from second quarter values. Lifecare then buys DrugX. What percentage of the third quarter profits was made by Lifecare?
A. 46.0%   C. 47.9%
B. 46.6%   D. 48.2%

Question 125:
Production costs are increased in the fourth quarter, resulting in all profits falling by a further 5%, despite an increase in sales. The information given in question 124 still applies. How much money does Lifecare make in this quarter?
A. £41,800   C. £43,490
B. £42,750   D. £47,002

SET 27

The chart below shows the cost of a variety of cars and optional extras. All prices are excluding 20% VAT, which must be paid by all customers.

Model	Price	Leather Seats	Sound System	Easy-Park Technology
Racer	£15,000	£395	£195	£395
Stuntman	£12,500	£345	£145	£295
Saloon	£21,500	£495	£245	£445
Pod	£18,000	£445	£395	£495

Question 126:

What is the total cost of the Stuntman, with all optional extras?

A. £15,942
B. £15,904
C. £15,894
D. £15,616

Question 127:

What is the price difference between the Saloon and the Pod (with all optional extras)?

A. £3,900
B. £4,020
C. £4,040
D. £4,100

Question 128:

What is the difference in price between the Racer (with no optional extras) and the Stuntman with all optional extras?

A. £2,040
B. £2,048
C. £2,058
D. £2,142

Question 129:

There is a 10% discount on the Racer and all its optional extras. What is the difference in price between the Pod with no optional extras and the Racer with all optional extras?

A. £4,236.20
B. £4,285.50
C. £4,336.20
D. £4,438.40

Question 130:

A final offer on the Saloon is 20% off, including all options. What is the difference in final price between the Saloon with Leather seats and Easy-Park technology and the Pod with only basic features?

A. £18.60
B. £37.20
C. £48.30
D. £57.60

SET 28

The graph below shows the total amount of CO_2 (in Tonnes) emitted by the country Aissur in each year from 2000 onwards.

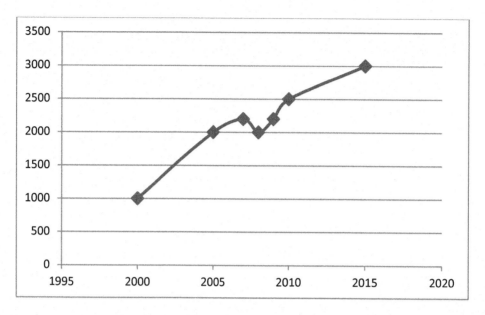

Question 131:
What was the rate of increase of CO_2 emissions between 2000 and 2005?

A. 250 Tonnes/year   B. 225 Tonnes/year   C. 200 Tonnes/year   D. 100 Tonnes/year

Question 132:
The economic crash of 2008 caused global CO_2 emissions to decrease due to a decrease in industrial output. How much less CO_2 was emitted in the year 2010 compared to if emissions had continued to rise at the same rate seen from 2000 to 2005.

A. 500 Tonnes   B. 750 Tonnes   C. 1,000 Tonnes   D. 2,500 Tonnes

Question 133:
What is the percentage increase in CO_2 emissions from 2005 to 2015?

A. 25%   B. 33%   C. 50%   D. 150%

Questions 134 – 135 require the following information:
In 2015, the government of Aissur voted on a new energy bill. The bill seeks to reduce the rate of CO_2 increase over the past 5 years by 50% over the next 5 years, and keep the increase at this level thereafter.

Question 134:
If the new energy bill is successful in meeting its aims, how much CO_2 will be saved by the end of 2020 relative to the 2010 – 2015 trend continuing?

A. 200 Tonnes   B. 250 Tonnes   C. 500 Tonnes   D. 750 Tonnes

Question 135:
What will the total CO_2 be in 2020 according to this new act?

A. 2,750 Tonnes   B. 3,000 Tonnes   C. 3,250 Tonnes   D. 3,500 Tonne

SET 29

The chart below shows the price per item for different styles of printing. The price is lower when larger orders are made, as shown in the table.

Type	1	10+	100+
Single sided black & white	£0.10	£0.07	£0.05
Single sided colour	£0.25	£0.20	£0.15
Double sided black & white	£0.15	£0.12	£0.10
Double sided colour	£0.45	£0.30	£0.25

Question 136:

What is the price per job of 74 single sided black & white sheets?

A. £3.70
B. £5.18
C. £5.24
D. £7.40

Question 137:

How many double-sided colour sheets can you buy for £100?

A. 222
B. 333
C. 400
D. 425

A 10% discount is offered for orders above 50 units, applying to the whole order. All other offers still apply.

Question 138:

What is the price of 150 units of double sided black & white?

A. £13.50
B. £15.50
C. £16.20
D. £20.25

Question 139:

Compared to buying 150 double sided black & white sheets individually, how many sheets worth (at the standard price for 1 sheet) is saved by buying in one transaction at the discounted price?

A. 65 sheets
B. 60 sheets
C. 53 sheets
D. 50 Sheets

Question 140:

What is the total cost of an order of double sided pages, with 227 requiring black and white printing and 34 requiring colour printing?

A. £22.38
B. £29.61
C. £32.90
D. £34.32

SET 30

4 sets of 300 volunteers take part in a clinical trial for a new drug, which is aimed at reducing the effects of asthma. The responses received are recorded below.

Group	Positive	Negative	No Effect
1	75%	20%	5%
2	65%	30%	5%
3	70%	15%	15%
4	55%	25%	20%

Question 141:

How many people reacted positively overall?

A. 135
B. 265
C. 523
D. 795

Question 142:

How many more people reacted negatively from set 2 compared to set 3?

A. 15
B. 33
C. 45
D. 56

Question 143:

What proportion of those tested overall reacted negatively?

A. 21%
B. 23%
C. 26%
D. 28%

After modifications to the drug, a new survey of 300 volunteers was taken. The results of this are shown below:

Group	Positive	Negative	No Effect
5	82%	15%	3%

Question 144:

What was the percentage increase in the success rate (i.e. the percentage of people reacting positively) in the 5th group compared to the first 4 groups?

A. 7.81%
B. 15.75%
C. 17.93%
D. 23.77%

Question 145:

Across all groups, including group 5, how many people reacted negatively to the drug?

A. 275   B. 315   C. 355   D. 380

SET 31

The graph below shows the total number of views for two rival local television dramas, The Last Chase and The Final Frontier, across the 4 yearly quarters in 2014.

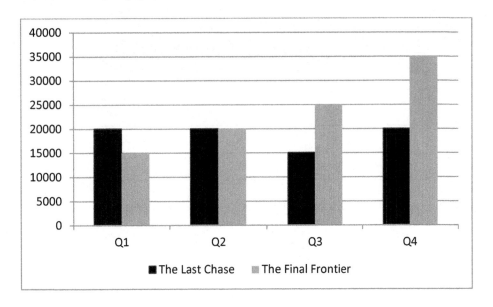

Question 146:
What is the difference between the total number of views of the Final Frontier and The Last Chase during 2014?
A. 10,000   B. 15,000   C. 20,000   D. 25,000

Question 147:
Broadcasters earn £2,500 from advertisements per 1,000 views. What is the difference in money earned through advertising between the two shows in 2014?
A. £45,000   B. £50,000   C. £55,000   D. £60,000

Question 148:
If the number of views of The Final Frontier continues to increase at same rate it did from Q1 – Q3 of 2014, how many views will it have during the final quarter of year 2015?
A. 50,000   B. 55,000   C. 60,000   D. 65,000

Question 149:
If the number of views of The Final Frontier continues to increase at same rate it did from Q1 – Q3 of 2014, how many views will it have in 2015 in total?
A. 180,000   B. 190,000   C. 200,000   D. 250,000

Question 150:
Under different circumstances, at the end of the third quarter of 2014, the broadcasters decide to terminate The Last Chase. As a result, half of The Last Chase's views instead transfer to The Final Frontier. How many views will The Final Frontier have at the end of the final quarter of 2014 under these circumstances?
A. 25,000   B. 35,000   C. 37,500   D. 45,000

SET 32

The table below shows the average time, in minutes, spent waiting for GP appointments by patients, according to a series of surveys from 2014. On average, 20% of patients who wait between 11 and 30 minutes and 40% of those who wait for more than 30 minutes register a complaint during a customer satisfaction survey. No patients who waited for 10 minutes or less registered complaints.

	0-10	11-30	30+	Survey size
England	60%	30%	10%	100,000
Scotland	55%	25%	20%	50,000
Wales	50%	25%	25%	25,000
Northern Ireland	60%	25%	15%	25,000

Question 151:
How many patients waited for less than half an hour for an appointment in Scotland?
A. 12,500
B. 27,500
C. 40,000
D. 45,000

Question 152:
What percentage of patients across the UK waited for more than half an hour for an appointment?
A. 10%
B. 15%
C. 20%
D. 25%

Question 153:
How many complaints are received from this survey at the end of the year?
A. 20,250
B. 21,500
C. 23,000
D. 24,250

Question 154:
What proportion of patients complained about waiting times by the end of the 2014 survey?
A. 12.1%
B. 11.5%
C. 11.0%
D. 10.7%

Question 155:
In January 2015, the government announced a target to reduce the number of patients waiting for longer than 30 minutes for an appointment by 50%, and by 25% for those waiting between 11-30 minutes. Proportionally, what will be the decrease in the number of complaints recorded by an identical survey at the end of 2015, if all targets are met?
A. 40%
B. 38%
C. 36%
D. 34%

SET 33

The graph below shows the price of crude oil in US Dollars during 2014:

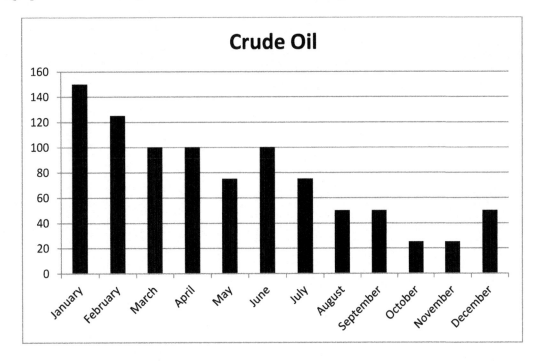

The total oil production, in millions of barrels per day, is shown on the graph below:

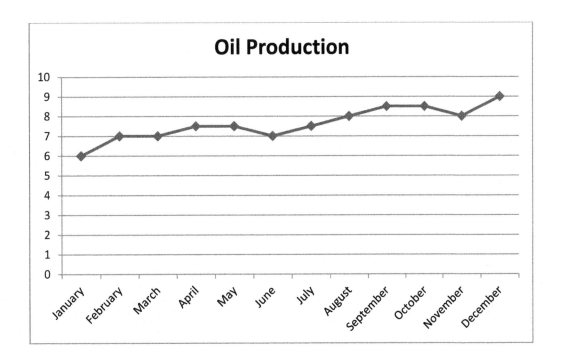

Question 156:
At what rate did the price of oil fall between January and March of 2014?

A. $16.70 per month
B. $20.00 per month
C. $22.70 per month
D. $25.00 per month

Question 157:
What was approximate total oil production in 2014?

A. 1,750 million barrels
B. 2,146 million barrels
C. 2,300 million barrels
D. 2,700 million barrels

Question 158:
How much did oil sales total in July 2014?

A. $0.56 Billion
B. $16.9 Billion
C. $17.4 Billion
D. $21.1 Billion

Question 159:
Oil prices have been falling due to a high supply. On average, the price of extraction & production of oil makes up 40% of the total price. The rest of the price is profit. How much profit was made from oil sales during June 2014?

A. $8.4 Billion
B. $12.6 Billion
C. $13.0 Billion
D. $21.0 Billion

Question 160:
Profit from oil extraction is 60% of the total sale price. This profit is split between the oil companies and the nation producing the oil in a ratio of 5:2. Of the oil company profits, 30% are used for corporation tax in the companies' home countries. Given that the overall sales value was $204 billion over the year, how much corporation tax was paid in 2014 (to 2 decimal places).

A. $26.23 Billion
B. $36.74 Billion
C. $43.71 Billion
D. $67.57 Billion

SET 34

The chart below shows the severity of asthma amongst a sample of 5 groups of 50 people of different ages. The average cost of asthma inhalers per patient is £50 per year. The population of the United Kingdom averaged 50 million during the period of interest. Children aged 0-5 years made up 7% of the population and children aged 5-10 years made up 10% of the population.

Children below the age of 10 who suffer from mild asthma have a half chance of developing respiratory problems in adult life. This figure is 90% for children below the age of 10 who suffer severe asthma. Children without asthma will not develop respiratory problems.

A review into doctors' practices concluded that between 1990 and 1995, 35% of mild asthma diagnoses of children between 0-10 were incorrect.

Age	No Asthma	Mild	Severe
0-5	80%	15%	5%
5-10	75%	20%	5%
10-21	85%	12%	3%
21-30	95%	3%	2%
30+	95%	4%	1%

Question 161:
How many people surveyed suffer from asthma?

A. 25   B. 35   C. 45   D. 55

Question 162:
What is the proportion of children surveyed who are likely to develop respiratory problems?

A. 13.00%   B. 13.25%   C. 13.50%   D. 14.25%

Question 163:
How many 0-10 year olds from the survey have been incorrectly diagnosed with asthma?

A. 6.1%   B. 6.9%   C. 7.4%   D. 8.0%

Question 164:
What proportion of children below the age of 10 who were correctly diagnosed with asthma will develop respiratory problems?

A. 9%   B. 10%   C. 11%   D. 12%

Question 165:
How much money was wasted on mistakenly prescribing medication to children who were wrongly diagnosed with asthma from 1990 to 1995?

A. £133 million   B. £157 million   C. £187 million   D. £255 million

SET 35

The following table shows data related to equity shares issued by five public sector companies on 1 March 2015.

Company	Number of equity shares (million)	Current market price Per share (£)	Percentage of equity share held by UK government
A	10	60	75%
B	20	50	50%
C	30	40	33.33%
D	40	30	25%
E	50	20	12.5%

Question 166:

If the government disinvested 50% of its stake in A at current market price, what (in £) is the amount of revenue generated by the government through the disinvestment?

A. 375 Million   B. 325 Million   C. 275 Million   D. 225 Million

Question 167:

If the government disinvested 25% of its stake in B at current market price, the amount of revenue generated by the government through the disinvestment would be (in £):

A. 125 Million   B. 150 Million   C. 175 Million   D. 200 Million

Question 168:

The government disinvested its entire stake in C at a price of £35 per share. What would have been the additional revenue generated by the government had it done the given disinvestment at the given market price?

A. £25 Million   B. £50 Million   C. £75 Million   D. £100 Million

Question 169:

If the share price of D fell to £25 on 2 March 2015, then what was the decline in the total value of D's shares held by the government from that of the previous day?

A. £25 Million
B. £50 Million
C. £75 Million
D. £100 Million

Question 170:

If the share price of E rose to £25 on 2 March 2015, then what was the increase in the total value of E's shares held by the government over that of the previous day (in £)?

A. £30.25 Million
B. £30.75 Million
C. £31.25 Million
D. £31.75 Million

Question 171:

Which of the following will fetch higher revenue for government?

A. Redeeming all its stock from company A.
B. Redeeming all its stock from company B.
C. Both of the above will fetch the same value.
D. None of the above.

SET 36

The table below shows the production of some agricultural crops in Harvestland in the years 2011-12 and the targets that were earlier set for that growing season.

Crop	Targeted production For 2011-12 (tonnes)	Actual production for 2011-12 (tonnes)	% Increase in production from 2010-11
Food grains	120	100	25
Oil seeds	60	50	25
Sugarcane	50	40	10
Cotton	40	30	20
Jute	25	20	25

Question 172:

The production of food grain (in tonnes) in 2010-11 was:

A. 40

B. 60

C. 80

D. 100

Question 173:

What was the difference in targeted production in 2011-12 and actual production in 2010-11 for oil seeds (in tonnes)?

A. 10

B. 20

C. 30

D. 40

Question 174:

How much more sugarcane should have been produced in order to meet the target in 2011-12 (in tonnes)?

A. 5

B. 10

C. 15

D. 20

Question 175:

What was the combined production of Cotton and Jute in year 2010-11 (in tonnes)?

A. 11

B. 21

C. 31

D. 41

Question 176:

How much more food grain was produced than oil seeds in 2010-11 (in TONNES)?

A. 10   B. 20   C. 30   D. 40

Question 177:

Cotton constituted what percentage of total crops in year 2011-12?

A. 10   B. 12.5   C. 15   D. 17.5   E. 30

SET 37

The table given below shows the sales volume of four products A, B, C and D manufactured by a company from January to April in the year 2014.

	January	February	March	April
Product A	9,500	10,250	10,500	11,000
Product B	6,500	7,000	7,250	7,500
Product C	3,500	3,750	4,000	4,250
Product D	2,500	3,100	3,500	4,000

Question 178:

In February, sale of product B constituted what percentage of total sales of all 4 products put together?
A. 26%
B. 27%
C. 28%
D. 29%

Question 179:

Which of the following products recorded maximum percentage increase from March to April?
A. Product A
B. Product B
C. Product C
D. Product D

Question 180:

In May 2014, the sales of product C witnessed an increase of 20% over the previous month. The sales of D were the same as those of C. What was the percentage increase in the sales of D in May relative to April?
A. 22.5 %
B. 25.0 %
C. 27.5 %
D. 30.0 %

Question 181:

By what percentage did the combined sales of product A and product C increase from January to April?
A. 17.0 %   B. 17.1 %   C. 17.2 %   D. 17.3 %

Question 182:

Assume a different scenario, that May 2015 witnessed a 20% growth in sales for products A and B, and a 30% growth in sales for products C and D over April values. What was the total sales value in May for all the products combined?
A. 32,925   B. 33,925   C. 34,925   D. 35,925

Question 183:

Assume a different scenario, that May 2015 witnessed 20% growth in sales of product A and 10% growth in sales for the other 3 products (B, C and D). Sales of A constituted what percentage of total sales in May 2015?

A. 40.25 %   B. 41.25 %   C. 42.25 %   D. 43.24 %

SET 38

A courier company uses three modes of transportation for delivering consignments – Road, Rail and Air. The following table shows the percentage distribution of the total number of consignments delivered, the revenue generated and the cost incurred, across the three modes of transportation in 2014.

Mode of transportation	Number of consignments (%)	Revenue (%)	Cost (%)
Rail	30	35	25
Road	45	20	25
Air	25	45	50

Question: 184

In 2012, the profit made by Courier Company was 30% of the total revenue. The company made a profit of £2.5 million. What was the total revenue?

A. £3.6 Million
B. £7.2 Million
C. £8.3 Million
D. £25 Million

Question 185:

In 2014, the cost per consignment was the lowest through which method?

A. Rail
B. Road
C. Air
D. Equal between road and rail

Question 186:

In 2014, the cost per consignment through rail was £5 and the revenue per consignment through rail was £20. What was the ratio of the total revenue through rail to the total cost through rail? Assume the number of consignments is equal to that given in the table.

A. 4:1   B. 5:1   C. 6:1   D. 7:1   E. 8:1

Question 187:

In 2013, the total costs of the company are £54,000. What is the total cost of air transportation in the year 2013?

A. £13,500
B. £17,000
C. £27,000
D. £32,000
E. More information needed

Question 188:

In 2014, if the total number of consignments delivered was 17,145, then what was total number of consignments delivered using rail and road?

A. 11,670   B. 11,974   C. 12,463   D. 12,859

SET 39

The following table provides partial information about the composition of three different alloys, A, B and C. Each of these alloys contains five different elements: Zinc, Tin, Lead, Copper and Nickel, and no other substances. An alloy, Alloy G, the composition of which is not given in the table, contains alloys A, B, C in the ratio 2:1:3. It is also known that in Alloy G, Tin, Lead and Copper are present in equal quantities.

Alloy	Zinc	Tin	Lead	Copper	Nickel
A	10%	40%			10%
B	25%	15%	50%	5%	5%
C	15%		20%		35%

Question 189:
Find the percentage of Lead in alloy A.
A. 8.33 %
B. 4.16 %
C. 2.70 %
D. 2.08 %

Question 190:
Find the percentage of Tin in alloy C.
A. 31.3 %
B. 15.8 %
C. 10.6 %
D. 7.9 %

Question 191:
An alloy X contains A, B and C in equal proportion. What is the percentage of Zinc in this alloy?
A. 12.50 %
B. 16.67 %
C. 25.00 %
D. 33.33 %

Question 192:
Find the percentage of Tin and Copper combined in alloy C.
A. 15 %
B. 20 %
C. 25 %
D. 30 %

Question 193:
Find the percentage of Tin in alloy G.
A. 11.11 %
B. 21.11 %
C. 31.11 %
D. 41.11 %

Question 194:
How many elements have exactly the same concentration in Alloy G?
A. One   B. Two   C. Three   D. Four

SET 40

The following table chart represents the number of people in the USA surveyed by CNN-Time in an opinion poll for "*The most influential person of the year 2001*". The number of people surveyed is 11,500.

Response	Percentage
Voted in favour of George Bush	39
Voted in favour of Donald Rumsfield	5
Voted in favour of Robert Guiliani	4
Voted in favour of Bill Clinton	2
Voted in favour of Lady Politicians	17
Non-respondents	33
Total	**100**

Question 195:
How many people voted in favour of Hillary Clinton, who received 60% of total votes polled for lady politicians?
A. 1,173   B. 1,223   C. 1,253   D. 1,273

Question 196:
If everyone who voted in favour of Robert Guiliani is a citizen of New York, then out of all the people surveyed, the number of citizens from New York is:
A. 460
B. 960
C. 1,040
D. Cannot be determined.

Question 197:
Out of the respondents, 20% are not US citizens. Given that only US citizens voted for George Bush, determine the percentage of US citizens who voted in favour of Bush.
A. 42.8%   B. 45.3%   C. 46.6%   D. 48.8%

Question 198:
Out of the total people surveyed, 40% are employees of the Federal Government and out of these 10% are in favour of Rumsfield. Find the number of people who are in favour of Rumsfield but are NOT employees of the Federal Government.
A. 105   B. 110   C. 115   D. 120

Question 199:
A mid-year survey has also been done on the same group of people. In that survey 16% of the people were in favour of Bill Clinton. Find the decrease in the number of people who voted in favour of Bill Clinton from mid-year survey to the actual survey.
A. 1,210   B. 1,410   C. 1,610   D. 1,810

Question 200:
A mid-year survey has also been done on the same group of people. In that survey 40% of the people were in favour of Bush. Find the decrease in the number of people who voted in favour of Bush from mid-year survey to the actual survey.
A. 115   B. 230   C. 460   D. 920

SET 41

Shop A sells merchandised mugs and also offers to print flyers at the prices shown below:

Number of flyers	Colour printing	Black and white printing
0-9	20p	5p
10-99	10p	4p
100+	5p	2p

Number of mugs	Basic mug cost per mug	Medium mug cost per mug	Premium mug cost per mug	Cost for logo/picture per mug
0-49	20p	30p	50p	5p
50-99	15p	26p	45p	4p
100-499	10p	20p	34p	3p
500+	7p	10p	20p	2p

Question 201:

Before Christmas, the shop offers an extra 15% discount. Customer A decides to buy 5 premium mugs and 4 basic mugs with a unique picture printed on the basic mugs. How much will this cost?

A. £2.50
B. £2.59
C. £2.76
D. £2.81
E. £2.98

Question 202:

Customer B needs 750 premium mugs, 130 basic mugs and 80 flyers, after Christmas. She wants a logo printed on the basic mugs. She wants black and white flyers. To the nearest pound, how much will she have to pay?

A. £160
B. £170
C. £172
D. £388
E. £395

Question 203:

Customer C wants to spend £250. How many medium mugs with a logo can she buy with this money if she also wants to buy 50 colour flyers?

A. 1200
B. 1225
C. 2041
D. 2452
E. 2500

Question 204:

The price of extra logo printed on the mug decreased by 1p, the price of premium mug decreased by 2p and the price of colour flyer decreased by 1p across all quantities in 2015. In percentage terms, how much cheaper would have been to order 70 medium mugs with logo and 150 colour flyers in 2015 instead of 2014?

A. 4%
B. 5%
C. 7%
D. 8%
E. 11%

Question 205:

The shop made a profit of £325,750 in 2015. This was a compound average growth rate of 1.2% between 2012 and 2015. What was the profit in 2012?

A. £188,513
B. £231,862
C. £314,299
D. £318,071
E. £362,059

SET 42

In St. Mary College, all students must have at least one device to interact with digital, interactive study materials. There are thirty students who have all three gadgets: Smartphone, tablet and laptop.

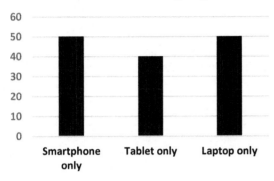

All students who only have one of the three gadgets

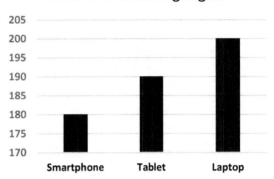

All students who have at least one of the three gadgets

Question 206:
How many students are studying at St. Mary College in total?
A. 325
B. 340
C. 345
D. 350
E. 360

Question 207:
How many students have both a tablet and smartphone but no laptop?
A. 40
B. 45
C. 50
D. 55
E. 65

Question 208:
How many more students have a smartphone than both a tablet and laptop?
A. 80
B. 85
C. 95
D. 100
E. 80

Question 209:
What percentage of all students have both a smartphone and a laptop?
A. 20.5%
B. 23.1%
C. 23.5%
D. 25.4%
E. 25.9%

Question 210:
Five more students come to St. Mary College. Three of the students have both a smartphone and a tablet. Two of the students have a smartphone only. What percentage of all students in the college have a smartphone now?
A. 34%
B. 45%
C. 54%
D. 55%
E. 60%

SET 43

Train Timetable [Cambridge to London Liverpool Street]

Cambridge	6.45	7.00	7.10	Then every 20 minutes	23.10
Whittlesford Parkway	7.25	7.40	7.50	Then every 20 minutes	23.50
Audley End	7.35	7.50	8.00	Then every 20 minutes	00.00
Bishops Stortford	7.42	7.57	8.07	Then every 20 minutes	00.07
Sawbridgeworth	7.53	8.08	8.18	Then every 20 minutes	00.18
Tottenham Hale	8.03	8.18	8.28	Then every 20 minutes	00.28
Liverpool Street	8.14	8.29	8.39	Then every 20 minutes	00.39

Question 211:
How many trains leave from Cambridge Station going to Liverpool Street between 2pm and 6.00pm?
A. 10
B. 12
C. 15
D. 18
E. 20

Question 212:
At 2.45pm, Antonia is at Sawbridgeworth station waiting for the next train to London Liverpool Street. What is the earliest time can she expect to arrive at Liverpool Street?
A. 3.01pm
B. 3.09pm
C. 3.19pm
D. 3.31pm
E. 3.51pm

Question 213:
How many trains run from Cambridge Station to Liverpool Street all day?
A. 46
B. 47
C. 50
D. 51
E. 56

Question 214:
It is 3pm and Mark wants to take the next train from Cambridge Station to Tottenham Hale for a 1 hour meeting straight at the station. What is the earliest time that Mark could schedule the meeting?
A. 4.25pm
B. 4.28pm
C. 4.45pm
D. 4.55pm
E. 5.02pm

Question 215:
If the distance between Audley End and Bishops Stortford is 10.5 miles. What is the speed of the train?
A. 60 miles/hour
B. 70 miles/hour
C. 80 miles/hour
D. 90 miles/hour
E. 100 miles/hour

SET 44

The list of the longest rivers in the UK by length in 2014.

River	Length (miles)	Drainage Area (square mile)	Average Discharge (m³/s)
River Severn	220	4,409	61.17
River Thames	215	4,994	65.8
River Trent	185	4,029	84
River Great Ouse	143	3,236	15.7
River Wye	134	1,597	3.4
River Ure	129	560	2.4
River Tay	117	431	2.3

Question 216:

What is the total length of the five longest rivers in the UK?

A. 897

B. 905

C. 1026

D. 1143

E. 1234

Question 217:

In 2015, the drainage area of the River Thames increased by 1345.5 square miles. In percentage terms, how much did the drainage area of the seven longest rivers in England increase in 2015 if the other rivers the same?

A. 2%

B. 7%

C. 17%

D. 93%

E. 97%

Question 218:

In percentage terms, how much bigger was the drainage area of the River Thames than the River Wye in 2014?

A. 213%   D. 345%

B. 276%   E. 425%

C. 313%

Question 219:

What was the difference in length between the two rivers with the highest and lowest average discharge?

A. 68 miles

B. 91 miles

C. 117 miles

D. 185 miles

E. 213 miles

Question 220:

What was the average drainage area of the six longest rivers in England in 2014?

A. 2750

B. 3138

C. 3210

D. 3455

E. 4235

SET 45

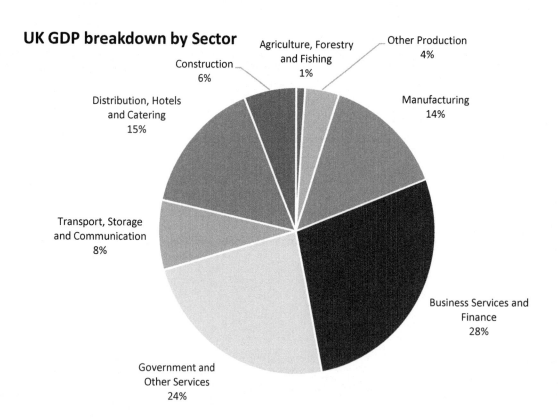

Scottish GDP breakdown by Sector

- Construction 7%
- Agriculture, Forestry and Fishing 1%
- Other Production 4%
- Distribution, Hotels and Catering 15%
- Manufacturing 14%
- Transport, Storage and Communication 7%
- Business Services and Finance 23%
- Government and Other Services 29%

UK GDP breakdown by Sector

- Construction 6%
- Agriculture, Forestry and Fishing 1%
- Other Production 4%
- Distribution, Hotels and Catering 15%
- Manufacturing 14%
- Transport, Storage and Communication 8%
- Business Services and Finance 28%
- Government and Other Services 24%

Question 221:

Which industry has the largest contribution to Scottish GDP?

A. Agriculture, Forestry and Fishing
B. Construction
C. Government and Other Services
D. Manufacturing
E. Other Production

Question 222:

In 2015, UK GDP increased by 5% overall to 4.2tn pounds. How much was the UK GDP in 2014?

A. 3.9tn
B. 4tn
C. 4.1tn
D. 4.2tn
E. 4.41tn

Question 223:

How much bigger is the contribution of Business Services and Finance to total UK GDP than the contribution of Manufacturing to Scottish GDP?

A. 3%
B. 4%
C. 5%
D. 7%
E. 14%

Question 224:

Scottish GDP was £1.2tn in 2014. What is the share of Agriculture, Forestry and Fishing in Scottish GDP in 2015 if there was an overall increase in GDP by 3%? Assume the percentage share of GDP for Agriculture, Forestry and Fishing does not change from 2014 to 2015.

A. £0.007tn
B. £0.012tn
C. £0.015tn
D. £0.022tn
E. £1.11tn

Question 225:

What are the top three contributing sectors to UK GDP?

A. Agriculture, Forestry and Fishing; Manufacturing; Construction
B. Government and Other Services; Manufacturing; Construction
C. Business Services and Finance; Government and Other Services; Manufacturing
D. Distribution, Hotels and Catering; Business Services and Finance; Government and Other Services
E. Distribution, Hotels and Catering; Business Services and Finance; Manufacturing

SET 46

Top-5 TV Shows	Viewers (millions)		% people watching show in Year 2	
	Year 1	Year 2	Females	Males
The Voice	3.4	4.1	5%	2%
Britain's Got Talent	5.2	5.6	4.5%	4%
Big Brother	1.3	1.4	9%	15%
Geordie shore	9.2	12.5	21%	19%
Fresh Meat	5	3	19%	5%

Question 226:

In percentage terms, how much did the most-watched TV show increase in the number of viewers between year 1 and year 2?

A. 29%

B. 33%

C. 36%

D. 41%

E. 42%

Question 227:

How many more viewers did The Voice and Britain's Got Talent have together in year 2 than in year 1?

A. 0.1 million

B. 0.2 million

C. 0.5 million

D. 0.7 million

E. 1.1 million

Question 228

Which is the least-watched TV show by males in year 2?

A. Big Brother

B. Fresh Meat

C. The Voice

D. Britain's Got Talent

E. Can't tell

Question 229:

In percentage terms, what was the total increase in the number of viewers across all TV shows from year 1 to year 2?

A. 5%

B. 10%

C. 15%

D. 20%

E. 25%

Question 230:

If the total population of females in year 2 was 30 million, how many males watched Geordie Shore in year 2?

A. 1.75 million

B. 2.38 million

C. 2.9 million

D. 6.2 million

E. 6.3 million

SET 47

Mobile Phone Plans

	Basic Plan Charges		Premium Plan Charges	
	Included minutes	Cost per additional minute	Included minutes	Cost per additional minute
Text Messages	250 free texts	15p per text	500 free texts	22p per text
Standard Calls	50	4p	75	6p
Mobiles (same network)	150	9p	500	15p
Mobiles (other networks)	100	15p	250	25p

Monthly fee for basic and premium plan fees: £45.29 and £47.89 respectively.

Question 231:

Each month, Claire sends 300 text messages and makes 75 mobile calls in the same network, each one minute on average. Which plan would be cheaper for Claire and by how much per month?

A. Basic plan by £4.90

B. Basic plan by £5.90

C. Premium plan by £3.90

D. Premium plan by £4.90

E. Premium plan by £5.90

Question 232:

Adam wants to spend maximum £60 per month on a mobile plan. Excluding any included minutes, what is the maximum number of additional minutes he can use if he has the premium plan?

A. 105 minutes

B. 145 minutes

C. 201 minutes

D. 245 minutes

E. 283 minutes

Question 233:

All basic plan charges increase by 1p; the basic plan fee remains unchanged. Andrew sends 45 texts, and uses 125 minutes in the same network and 325 minutes in other networks on average. How much is he worse off?

A. 0p

B. 75p

C. £1.55

D. £2.05

E. £2.25

Question 234:

Daisy does not send any text messages ever. She makes 800 minutes of calls in the same network. Which plan would be cheaper for Daisy and by how much?

A. Basic plan by £2.10

B. Premium plan by £5.75

C. Premium plan by £9.33

D. Premium plan by £10.90

E. Premium plan by £11.89

Question 235:

Kevin sends only text messages, 450 per month on average. If there were a 15% price increase in the monthly fee for both basic and premium plans, how much would Kevin save by changing to premium plan now?

A. £23.43

B. £25.01

C. £27.01

D. £27.10

E. £28.91

SET 48

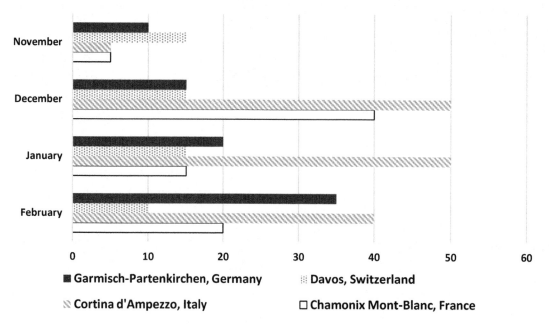

2015 Winter Snowfall (cm)

- ■ Garmisch-Partenkirchen, Germany
- ▦ Davos, Switzerland
- ▨ Cortina d'Ampezzo, Italy
- ☐ Chamonix Mont-Blanc, France

Question 236:

What was the mean monthly snowfall in cm across Davos and Chamonix Mont-Blanc during winter in 2015?

A. 12.325 cm
B. 16.875 cm
C. 19.738 cm

D. 26.842 cm
E. 43.123 cm

Question 237:

During Winter 2015, where was the average monthly snowfall the highest?

A. Davos
B. Chamonix Mont-Blanc
C. Cortina d'Ampezzo

D. Garmisch Partenkirchen
E. Can't tell

Question 238:

In percentage terms, how much more snow fell in December than in February overall?

A. 8%
B. 12%

C. 14%
D. 20%

E. 22%

Question 239:

In November 2014, 30cm snow fell in the four areas together. In percentage terms, how much more snow fell in November 2015 in the four areas together?

A. 5%
B. 17%
C. 19%

D. 24%
E. 29%

Question 240:

How much snow fell in Cortina d' Ampezzo and Garmisch Partenkirchen in November and February together?

A. 20 cm
B. 30 cm
C. 60 cm

D. 90 cm
E. 145 cm

SET 49

Please find below Kevin's expenses for December 2015.

Date	Client	Detail of expense	Cost
01.12.15	HSBC	Mileage to and from presentation in Cambridge	£16.80
01.12.15	Soros Fund Management	Single ticket - Train journey to meeting in London	£15.40
13.12.15	HSBC	Breakfast with client	£35.90
16.12.15	Black Rock	Return ticket – Train journey to meeting in London	£20.00
20.12.15	MKB	Lunch with client	£49.50

Mileage paid at £0.25 per mile for the first 100 miles each month and £0.10 thereafter

Question 241:

The greatest proportion of December expenses related to which client?

A. HSBC

B. Soros Fund Management

C. Black Rock

D. MKB

E. Can't Tell

Question 242:

What is the total expense in December?

A. £121.50

B. £137.60

C. £142.45

D. £146.50

E. £210.40

Question 243:

In percentage terms, how much more did Kevin spend on meals than on train tickets?

A. 35%

B. 41%

C. 59%

D. 149%

E. 141%

Question 244:

How many miles did Kevin travel by his car in December?

A. 50.5 miles

B. 67.2 miles

C. 67.9 miles

D. 78.4 miles

E. 112.6 miles

Question 245:

The company decides to change its policy from 2016 and only 75% of travel expenses and 90% of accommodation and meal expenses will be reimbursed. How much money would Kevin have lost in December if the new policy had been implemented already in 2015?

A. £8.54

B. £14.22

C. £13.05

D. £21.59

E. £39.15

SET 50

The table below shows the cost of jet ski renting. There are four different jet skies: Alpha, Beta, Gamma and Delta. The deposit is non-refundable.

Type	Deposit	Cost per hour	Total cost for first hour
Alpha	---	£30	£50
Beta	£20	£90	£110
Gamma	£25	£115	£140
Delta	£100	£150	£250

Question 246:

How much is the deposit for Alpha jet ski?

A. £20

B. £30

C. £40

D. £50

E. £60

Question 247:

What is the difference between the total cost of renting a Beta and a Gamma jet ski for 6 hours each?

A. £150

B. £155

C. £200

D. £205

E. £215

Question 248:

Aron has £500 for jet ski renting. Which jet skis can he afford to rent for an hour?

A. Alpha only

B. Alpha and Beta and Gamma

C. Alpha and Beta and Gamma and Delta

D. Beta only

E. Gamma only

Question 249:

Andrew and Thomas want to rent Alpha and Gamma jet ski for three hours. How much will they pay in total?

A. £360

B. £480

C. £660

D. £690

E. £960

Question 250:

If the deposit for Delta jet ski is increased by 5% on Sundays, what will be the total cost of renting a Delta jet ski for 8 hours on a Sunday?

A. £990

B. £1200

C. £1305

D. £1605

E. £1900

SET 51

Flight tickets to various regions from the UK

	In The Air	Good Fly	Take Me There	Around the World
North America	250£	290£	560$	360 €
South America	190£	275£	370$	300 €
Europe	80£	100£	210$	110 €
East Asia	290£	280£	570$	400 €
Australia	290£	300£	610$	450 €

Assumed exchange rate is 1£ = 2$ = 1.5€

Question 251:

What is the price of the cheapest offer to East Asia in pounds (£)?

A. £238

B. £252

C. £267

D. £276

E. £280

Question 252:

On average how much more expensive is it to choose a non-European destination with Good Fly than to choose a European one? (rounded up to one digit)

A. £185.2

B. £186.3

C. £186.9

D. £189.8

E. £191.0

Question 253:

Take Me There decides to offer a 50$ discount on every travel to the Americas. In percentages, how much more expensive is the discounted ticket to North America than the original price offered by In The Air?

A. 2%

B. 4%

C. 5%

D. 7%

E. 11%

Question 254:

Around The World only sells 10 tickets to Australia and 5 tickets to South America.

Good Fly only sells 25 tickets to Europe and 12 tickets to Australia. Rounded to two decimals, what is the difference between the total revenues of these two companies, in percentage of the higher revenue?

A. 33.33%

B. 34.86%

C. 34.43%

D. 35.35%

E. 36.36%

Question 255:

The £ to $ exchange rate changes in a way that £1 = 2.5$. In pounds, what is the difference between the original and the new ticket price of Take Me There to Australia?

A. £55   B. £58   C. £61   D. £64   E. £75

SET 52

Indian GDP

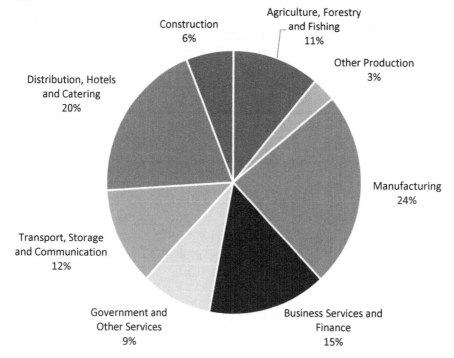

UK GDP

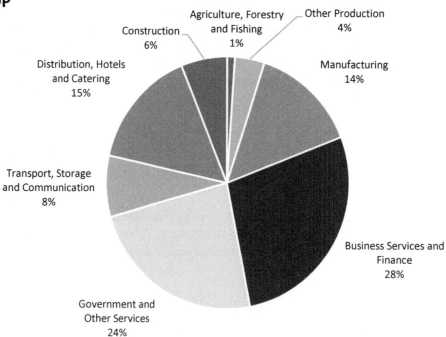

Question 256:

Which industry has the largest contribution to Indian GDP?

A. Agriculture, Forestry and Fishing
B. Construction
C. Government and Other Services
D. Manufacturing
E. Other Production

Question 257:

Expressing 'Government and Other Services' as a percentage of 'Business Services and Finance', what is the difference between these ratios in India and the UK?

A. 10%
B. 12%
C. 26%
D. 34%
E. 67%

Question 258:

What are the top **three** contributing sectors to Indian GDP?

A. Business Services Finance; Distribution, Hotels and Catering; Manufacturing
B. Agriculture, Forestry and Fishing; Business Services Finance; Manufacturing
C. Distribution, Hotels and Catering; Government and Other Services; Business Services and Finance
D. Construction; Manufacturing; Transportation, Storage and Communication
E. Distribution, Hotels and Catering

Question 259:

If the Indian GDP was £2 trillion and the UK GDP was £4.2 trillion in 2014. How much more did the bottom two performing sectors contribute to the UK GDP than to the Indian GDP?

A. £10 Billion
B. £15 Billion
C. £21 Billion
D. £30 Billion
E. £40 Billion

Question 260:

If the Indian GDP was £2 trillion and the UK GDP was £4.2 trillion in 2014. In monetary terms which sector was the largest?

A. Business Services and Finance in the India with £400 Billion
B. Business Services and Finance in the UK with £400 Billion
C. Manufacturing in India with £1.176 trillion
D. Business Services and Finance in the UK with £1.176 trillion
E. Business Services and Finance in the UK with £2 trillion

SET 53

The table below shows the average value of 1 Japanese Yen (JPY).

	2010	2011	2012	2013
GBP	0.021	0.020	0.019	0.022
USD	0.010	0.013	0.013	0.015
EUR	0.015	0.016	0.015	0.015
CAD	0.123	0.115	0.119	0.125

Question 261:

How much USD could you get for 1500 Yen in 2011?

A. USD 19.50

B. USD 21.00

C. USD 21.90

D. USD 25.50

E. USD 100.50

Question 262:

What was the GBP/USD average exchange rate in 2013 based on the information available?

A. 0.85

B. 0.92

C. 1.26

D. 1.47

E. 1.54

Question 263:

In percentage terms, what was the increase in JPY/CAD exchange rate between 2011 and 2012?

A. 3.5%

B. 3.9%

C. 4.5%

D. 5.1%

E. 11.5%

Question 264:

How much more JPY would I have received for USD 1300 in 2010 than in 2013?

A. JPY 43,333

B. JPY 45,000

C. JPY 45,225

D. JPY 46,310

E. JPY 47,450

Question 265:

Which exchange rate has been the least volatile between 2010 and 2013?

A. CAD

B. USD

C. EUR

D. JPY

E. GBP

SET 54

The table below shows changes in car use and population in four American cities between 2009 and 2014.

		Boston	Chicago	Denver	El Monte
2009	Population	4 million	3.5 million	2 million	0.8 million
	Number of cars	235,675	345,526	231,456	54,000
2014	Population	4.4 million	3.3 million	2.1 million	1.5 million
	Number of cars	542,000	350,685	249,990	62,044

Question 266:

In which city was the population growth the largest between 2009 and 2014?

A. Boston
B. Chicago
C. Denver
D. El Monte
E. Can't Tell

Question 267:

In which city was the growth in the number of cars used the largest between 2009 and 2014?

A. Boston
B. Chicago
C. Denver
D. El Monte
E. Can't Tell

Question 268:

In which city was the number of cars per person the lowest in 2014?

A. Boston
B. Chicago
C. Denver
D. El Monte
E. Can't Tell

Question 269:

In which city was the change in the number of cars per person the largest between 2009 and 2014?

A. Boston
B. Chicago
C. Denver
D. El Monte
E. Can't Tell

Question 270:

What is the difference between the number of cars per person in 2009 and 2014 in Boston?

A. 0.0064
B. 0.035
C. 0.054
D. 0.064
E. 0.065

SET 55

The table shows the change in tax rates and bands from tax year 2014-15 to 2015-16.
Family: Adam, Lewis, Courtney and Bruno

Income Tax Bands	2014-15 Rate	2015-16 Rate	2014-15 (GBP)	2015-16 (GBP)
Starting rate	15%	10%	$\leq 2,450$	$\leq 2,730$
Basic rate	25%	25%	$2,450 <$ to $\geq 33,500$	$2,731 <$ to $\geq 37,000$
Higher rate	40%	40%	$33,500 <$	$37,000 <$

Question 271:

Adam earned £37,000 in 2014-15. How much income was deducted from his salary during that year?

A. £4560

B. £6730

C. £8129.50

D. £9110.75

E. £9530

Question 272:

Bruno and Lewis each have a part-time job at a local pub and each earned £7,000 per year in 2014-15. What is the difference between their annual incomes after income tax has been deducted?

A. £0

B. £50

C. £176

D. £450

E. £745

Question 273:

Courtney received £42,000 in the tax year 2014-15. What was her average monthly income tax deduction?

A. £878.98

B. £898.98

C. £910.15

D. £960.83

E. £1024.58

Question 274:

Adam earned a performance bonus of £4,000 in 2014-15, increasing his salary from £37,000 to £41.000. By how much did his income tax change from what it would have been for this year without the performance bonus?

A. £1200

B. £1600

C. £1800

D. £2100

E. £3400

Question 275:

How much did the starting rate upper bound change from 2014-15 to 2015-16?

A. £80 decrease

B. £150 decrease

C. £280 decrease

D. £150 increase

E. £280 increase

SET 56

The table below shows the annual summary for performance evaluation of four partners at a management consulting firm. The bonus earned is calculated by multiplying the total sales by the bonus rate.

Partner	Projects Acquired (No.)	Total Sales	Bonus	Customer Satisfaction Rate
Adam	5	£45,000,000	1%	97%
John	2	£150,453,000	5%	90%
Richard	3	£180,000,000	4%	75%
Daniel	7	£654,150,000	10%	25%

Question 276:
Who received the highest bonus?
A. Adam
B. John
C. Richard
D. Daniel
E. Can't Tell

Question 277:
What is the average sales per project generated by Daniel?
A. £83,755,000
B. £93,450,000
C. £96,345,000
D. £99,950,000
E. £100,145,000

Question 278:
How much bonus did Richard and Adam receive together?
A. £5,500,000
B. £6,000,000
C. £6,350,000
D. £7,250,000
E. £7,650,000

Question 279:
Whose customer satisfaction rate was the highest?
A. Adam
B. John
C. Richard
D. Daniel
E. Can't Tell

Question 280:
How much was the total sales generated by the four partners together?
A. £1,029,603,000
B. £1,135,150,000
C. £1,529,540,000
D. £1,529,110,000
E. £2,130,042,000

SET 57

The table below shows the daily share price movements of four UK companies:

Name	Current Share Price (in pence)	Percentage Change from Previous Day	High	Low	Volume
HSBC	25.432	-0.03%	27.000	25.123	7,345,321
BP	286.123	+4.5%	286.456	284.567	4,431,748
GSK	134.432	+0.13%	142.511	131.678	1,125,469
British Land	54.923	+1.11%	54.934	54.914	4,999,432

Market Capitalisation is calculated as: number of shares outstanding (volume) multiplied by share price

Question 281:
What was the share price of GSK yesterday?
A. £132.873
B. £134.157
C. £134.257
D. £134.345
E. £135.012

Question 282:
What was difference between the daily highest and lowest price of British Land?
A. £0
B. £0.01
C. £0.02
D. £0.025
E. £0.3

Question 283:
What was the market capitalisation of HSBC yesterday assuming that the volume is unchanged?
A. £186,250,178.618
B. £186,350,178.618
C. £186,450,178.618
D. £186,650,178.618
E. £186,862,262.351

Question 284:
Which company's actual share price changed the most from yesterday?
A. HSBC
B. British Land
C. BP
D. GSK
E. Can't Tell

Question 285:
What is the difference between the latest market capitalisation of BP and British Land?
A. 993,441,229.268
B. 993,941,234.268
C. 994,114,112.268
D. 994,241,299.268
E. 1,113,345,891.268

SET 58

The table below is a summary of students who signed up for the following courses at St. Mary Grammar School:

Courses	Women	Men
Psychology	10	6
Maths	8	7
Physics	10	15
Programming	4	5
Literature	12	8
History	7	7

Students can take more than one course.

Question 286:
For which course is the ratio of women and men most similar to that of Psychology?
A. Mathematics
B. Physics
C. Programming
D. Literature
E. History

Question 287:
For which course is the ratio of women and men most similar to that of Physics?
A. Mathematics
B. Psychology
C. Programming
D. Literature
E. History

Question 288:
What is the total number of women in St. Mary Grammar School?
A. 34
B. 51
C. 64
D. 145
E. Can't Tell

Question 289:
What is the total proportion of women to men?
A. 0.25
B. 0.5
C. 0.96
D. 1
E. Cannot Say

Question 290:
If three new students arrive at St. Mary Grammar School and they are all women studying Psychology. What is the change in the ratio of women to men studying Psychology?
A) 0.1
B) 0.3
C) 0.5
D) 0.7
E) 1.2

SET 59

This graph below shows the employment statistics (in percentages) of the men and women living in the UK in September 2015. A person is considered being in employment if they are shown as employed, self-employed, or "employed other".

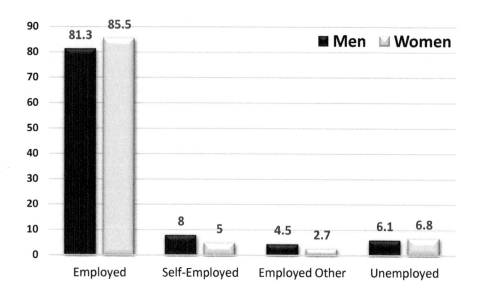

Question 291:
The difference between the percentage of women and men in employment is:

A. 0.6   B. 0.7   C. 1.2   D. 3.2   E. 4.2

Question 292:
The proportion of men to women self-employed or unemployed is:

A. 0.1   D. 1.2
B. 0.5   E. 1.7
C. 0.9

Question 293:
If there are overall 21 million women in employment in the UK how many women are self-employed?

A. 0.59 million   D. 1.07 million
B. 0.86 million   E. 1.13 million
C. 0.93 million

Question 294:
If there are 31 million men and 32 million women living in the UK, how many more women are unemployed than men?

A. 285,000   D. 735,000
B. 423,000   E. 810,000
C. 659,500

Question 295:
What is the proportion of employed to unemployed women in the UK?

A. 2.87   D. 13.7
B. 5.43   E. 15.91
C. 8.51

SET 60

This chart shows the expenses of the Jones household for this month. Overall they spent 360 pounds.

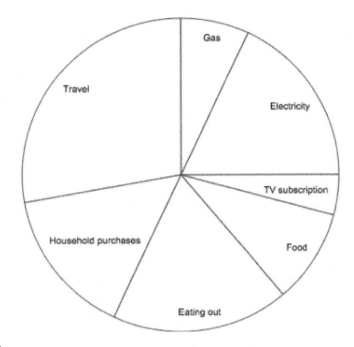

Question 296:
How much was their energy cost (electricity and gas) this month?
A. 65 pounds
B. 85 pounds
C. 90 pounds
D. 155 pounds
E. 200 pounds

Question 297:
What is the ratio of household purchases, travel and gas costs to energy costs (electricity and gas)?
A. 0.5:1
B. 1:1
C. 1.5:1
D. 2:1
E. 3:1

Question 298:
Which cost item was the greatest this month?
A. Gas
B. Travel
C. Electricity
D. Household purchases
E. Energy

Question 299:
If the total cost of household purchases, travel and gas costs was 180 pounds. How much were the energy costs in total this month?
A. 80 pounds
B. 87.5 pounds
C. 89.5 pounds
D. 90 pounds
E. 95 pounds

Question 300:
Which was the smallest cost item this month?
A. Gas
B. Travel
C. Electricity
D. Household purchases
E. TV subscription

Abstract Reasoning

The Basics

The abstract reasoning section of the UKCAT will test your ability to think beyond the information that is readily available to you in form of the information provided by the question. The idea behind this section of the paper is to test how well the candidate is able to respond to questions that may go beyond the scope of their knowledge or require them to apply their existing knowledge in an unusual way. This is thought to be helpful in determining how well a student will be able to interpret information such as scans, X-rays or other test results as a clinician.

This section of the test examines pattern recognition and the logical approach to a series of symbols in order to match symbols to one group or another. There are a number of different question types, but all require one key skill – the ability to recognise patterns in a set of shapes

In this section of the UKCAT, you have to answer 55 questions in only 13 minutes (with one additional minute to read instructions). Thus, it is mathematically the most time pressured section of the UKCAT. But in terms of timing, think of it in terms of the image sets. There are multiple questions per image set. Since the main investment in time is in figuring out the pattern, you have a greater proportion of the time to spend on the first question in each set. Then all subsequent questions in that set will be easy and quick to answer. By far the hardest task is deducing the rules – once you have them, matching the options to the correct set is straightforward. Therefore as a rule of thumb, if the image set has 5 questions on it you have about 60 seconds to work out the pattern. Then, match the options to the set they each belong in using the remaining time allocation.

Timings

As with the rest of the test, you have to keep an eye on the time to keep track of how you're doing. Make sure that you stay within your time limit of 78 seconds for each block of 5 questions, and quicker for data sets with fewer questions. When divided up to 15 seconds per question it might not seem like a lot, but actually given the format of the questions you will begin to realise that it is enough. As mentioned above, for the 5-question sets keep ticking on at a steady rate of about 55–60 seconds to find the rule then about 18–23 seconds to decide which set the 5 different options fit into. Your **practice will increase both your speed and overall likelihood of finding the rule**, but despite thorough preparation you still may fail to spot the pattern. If you can't see it, don't despair – simply make reasonable guesses (you have a 33% success rate by chance alone), flag for review (in case you have spare time at the end to check back and have another go) and move swiftly on.

> *Top tip!* Give yourself plenty of time to systematically work out the rules. Once you have found them, answering the questions will be quick.

Pattern Recognition

By far the most important ability in this section is to correctly identify patterns, as regardless of question style the matching process is straightforward once you have identified the rules. Some people are naturally better at this than others. You might be the sort of person who sees these patterns easily and can quickly put a name to the rule, or you might be the sort of person who finds it takes them more time and effort to work out what's going on. In reality, everyone lies at a different position on this scale, but one thing is certain. **You can improve your speed and accuracy on this section by having a methodical system** that can be repeated and applied to all shape-sets. One such system is the **NSPCC** system. This provides a logical structure for working through each set of images and looking for different components of a possible pattern.

In this system, the letters stand for:

<u>N</u>umber ➔ <u>S</u>ize ➔ <u>P</u>osition ➔ <u>C</u>olour ➔ <u>C</u>onformation

Using this system, you consider each of the following aspects of the images in sequence, looking each time for commonly used patterns. We recommend this because it begins by looking for the simplest and most common potential patterns – if they are present, you are sure to get the pattern quickly and easily. If the first few patterns you look for are not present, then you look further on in the sequence to check for harder and less commonly tested patterns until you arrive at the answer.

Practice

It's very important to practice for this section as the style of questions are unlikely to be familiar. **Practicing well gives you three key advantages**. Firstly, you get used to the types of patterns which are likely to be asked in the real exam. This makes it more likely you will spot the patterns quickly as you will have seen them before, and it also trains up your implicit recognition system, meaning that if you take an "educated guess" you are more likely to be right. Secondly, it gives you practice implementing a pattern recognition system, like **NSPCC**. With practice you will become better at using the system, and therefore quicker and more accurate overall. Thirdly, you will gain a feeling for the time it takes to answer different types of question. This will allow you to better plan your time on the day, making the most out of every second you have.

Guessing

If you practice well you shouldn't have to guess very many questions, but it might be necessary if you just can't figure out a pattern. Guessing in this section of the test has a reasonable probability of success. Since there are only three or four options per question there is a 25-33% chance of guessing any one question correctly, and if all 5 questions in a data set are guessed then there is and 87% chance of gaining at least one mark from the set – better than in any other section of the UKCAT. However there is more to it than that.

Whilst the best way to answer these questions correctly is to formally deduce the rule and apply it (using a pattern recognition system like **NSPCC** helps here), humans *do* have an innate instinct for pattern recognition. This innate instinct is not necessarily right and can lead you astray, but in a quick guessing situation, it can be applied cleverly to boost your chances of guessing correctly. That is to say, in some questions the overall look of the image will feel as though it should be placed in a particular set – you wouldn't be able to say exactly why, but to you it would look much more like one group than the other. Learning to harness this power can help give you a much better guessing accuracy. Below is a simple example to demonstrate:

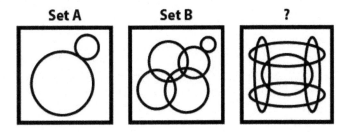

By quickly looking at the images provided for Set A and B, you get a general feel for what the contents are like. If asked which set the question image belongs to, you might be tempted to say Set B – it kind of looks more complicated and cluttered then Set A. It just looks more like the image. Now whilst this reasoning does not provide a comprehensive rule, in instances like this it can lead you to the correct answer quickly – even if you don't properly identify the underlying pattern.

Using the NSPCC system

The **NSPCC** system is a good way of working through possible patterns in a structured way. Whilst no structured response can be perfect, this system will solve over 90% of patterns quicker and more reliably than by trusting intuition alone.

Using the system, it is important that you are thorough. Sometimes the pattern can be subtle and you could easily miss out on it if not taking care. You have to examine the details closely: count corners, sides of shapes, check where they are etc. **You shouldn't be looking AT the shapes; you should be looking FOR patterns**. You should be working quickly, checking one thing, and if it's not that checking the next item in your list until you find the rule. Now to look at the system step by step.....

Number

Looking at "number" is about *counting* as many things as possible. How many dots? How many squares? How many sides? How many corners? How many right angles? Also have a look at how many different types of shapes you can find in the frames. Sometimes you might find one type of shape only in Set A and not in Set B for example. A good rule of thumb: block arrows have 7 sides, so don't count all the sides individually every time you see one!

Size

It is quite common to find patterns in the size of the shapes. Is one shape always bigger than the others? Is there always a big shape in the centre, or in the corner? Are there smaller shapes inside larger ones?

Position

Look for patterns in where shapes are positioned. You might, for instance, always find a square in the top right corner in one set and a circle in the top right corner of the other set. Look for Look also for touching and overlap of shapes – when you see this, make careful note of the type of contact. Is it tangential? Does it cut the shape in equal pieces, or is it off centre? Is there a certain shape that always makes this contact?

Colour

The shading of different shapes can constitute a pattern. Whilst this is often the easiest pattern to spot, it takes its place lower down in the system as it can often be a distracter. Most diagrams contain some amount of shading, but only occasionally is the primary pattern centred upon this. Look for shapes that are always shaded. Are all triangles black in one set and all circles black in the other, for example? On the other hand, are some shapes never shaded?

Conformation

These are the hardest patterns to spot, as they are the more complex patterns that can't be found by looking at the more geometric aspects. Conformation describes the pattern by which the shapes are arranged within the box – so you have to take a step back and look at the box as a whole in order to spot them. Look for patterns to the arrangement, like shapes arranged in a horizontal, vertical or diagonal line. Look also for the influence of one shape on another. For example, the presence of a white circle might signal a 90 degrees clockwise rotation of one shape and the presence of a black circle might signal a 90 degrees anticlockwise rotation, for example. When there are arrows, look at where they are pointing: are they all pointing in the same direction or at the same thing? You're looking for second order patterns, how things change based upon other aspects of the image.

Question Answering Strategy

There are four styles of question in this section, but all of these require the same pattern-recognition skills.

The **first style** of questions is the original style of question that used to be the only style in this section of the UKCAT. It also tends to be the style that accounts for the majority of the questions in this section, however this is not an official rule and it will not necessarily be the same this year. In this type of question, you are provided with two sets of six shapes, Set A and Set B. All of the images within each set are linked to each other by a common rule, but the rule must be different for Set A and Set B. The task is to identify the rules for each set, then for the 5 options you need to decide where they belong. If an option follows the same rule as the shapes in Set A, then it belongs in Set A, and likewise for Set B. If the image obeys neither the rule for Set A nor the rule for Set B, then it is correct to say it belongs in neither set and you choose the "neither" option. Approach this style of question by spending the majority of the time deducing the rule by using a system like **NSPCC**. Then when you have decided on the rule for each set, work through the options, selecting which set each fits best. If you don't figure out the rule in time – don't worry. It is expected that you won't work out them all in the time pressure of the exam. In that case, simply use an educated guessing strategy to give yourself a good chance of picking up some marks and then move on.

> *Top tip!* If a shape fits the rules for **both sets**, then the correct response is always "**neither**"

In the **second style** of question, you are provided with a single sequence of four shapes. They should be read as a sequence from left to right. You are then asked to choose the next shape in the sequence out of four options. This style of question is normally quicker to answer as it is more intuitive. In addition, you have three different transition points that you can compare to each other to help deduce whether you have correctly identified the rule. To answer this style of question well, start by scanning quickly across all four shapes – this gives you a general understanding of what is happening in the sequence. Then, focus on the element that is changing and apply your system to find out exactly *how* it is changing. Once again, if you're struggling, you can probably use your intuition to make a decent educated guess and move on. Consider flagging for review so you know where to focus your efforts if you have any additional time at the end.

The **third style** of question is also a sequence style of question. You are provided with two shapes with a rule linking them. The rubric states shape one **is to** shape two by this rule. Then you are provided with shape three, and you have to apply this rule to find out how the rule transforms shape three into shape four (you are given four options). Once again, the key to accuracy is in deducing the rule that links the shapes. Focus on the same elements in shape one and shape two and notice any changes that have taken place. This style of question has a more straightforward strategy as there are many fewer options to examine; as you can directly compare the two boxes you can usually deduce the rule without needing to use a system, but remember it is always there if needed. Be certain to check the rule applies to *all* elements in the boxes, otherwise you will need to revise the rule to account for the box as a whole rather than just one or two of the elements. Once you are satisfied with the change, apply the rule to the next shape and select the answer. Ideally you should imagine what the answer is *before* focusing on the options, otherwise you could be biased by a similar but incorrect option, but if you are struggling to do this then use your intuition to select the option that feels as though it is the best fit.

The **fourth style** of question is very similar to the first and original style. You are provided with two sets of six shapes, Set A and Set B, each linked by a common rule. Once again apply the same system to deduce the rule for each set. The only difference comes when it is time to select your answer. Instead of being asked which set a shape fits into, you are provided with four options and asked to select the one that fits into one set or the other. So once you know the rule, test the options to see which one fits the set you are asked for.

Example 1

Start by applying the NSPCC system

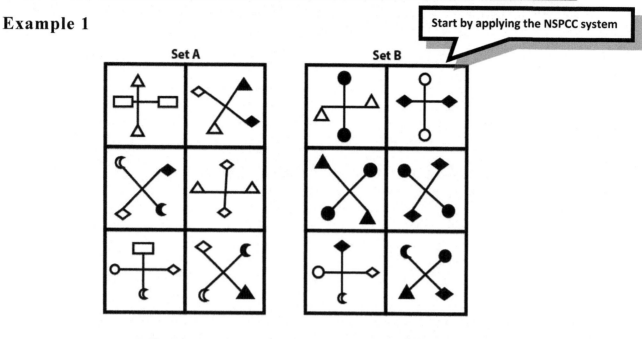

Set A   Set B

Test shape 1   Test shape 2

Start by applying the NSPCC system. Number: count the number of shapes, sides, angles and so on looking for a pattern (I can't see one). Size: are all shapes the same size (yes they are). Position: is there a pattern to where certain shapes are (not obviously). Colour: Is there a pattern to the shading (yes, the shading is dependent upon the shape of the cross). If you didn't get that, look back now to identify what the pattern is before reading on.

We can use our observations to devise the following rules:
Set A: (+)-shaped crosses have four white shapes and (X)-shaped crosses have two white and two black shapes.
Set B: (+)-shaped crosses have two white and two black shapes and (X)-shaped crosses have four black shapes.

Applying these rules tells us Test Shape 1 belongs to Set B and Test Shape 2 belongs to Set A.

Example 2

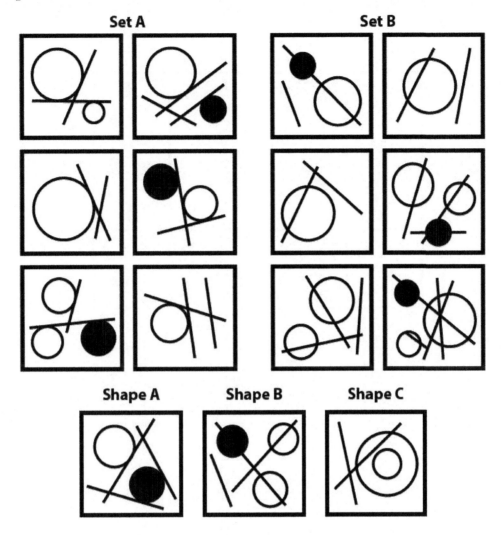

Start by applying the NSPCC system. Once again there is no pattern for number or size, but when reaching position you should note that shapes are touching and intersecting. Look more closely at this to work out what the pattern is.

It turns out that the rules for the set are as follows:

Set A: Every circle is touched tangentially by at least one line

Set B: Every circle is intersected by at least one line

Applying these rules tells us Shape 1 belongs to Set A, Shape 2 belongs to Set B and Shape 3 belongs to neither set, as the middle circle is neither touched tangentially nor intersected by any line.

A Final Word

This section is all about pattern recognition. The more you see the better you will become. Once you're familiar with the main types of patterns which come up, you'll be able to solve the majority of questions without difficulty. Remember that **you're looking to identify a rule** for each set of boxes, something which links them all together. Then, you can decide which set each question item fits into (or indeed neither). Start using the **NSPCC** system, then practice makes perfect!

Abstract Reasoning Questions

SET 1

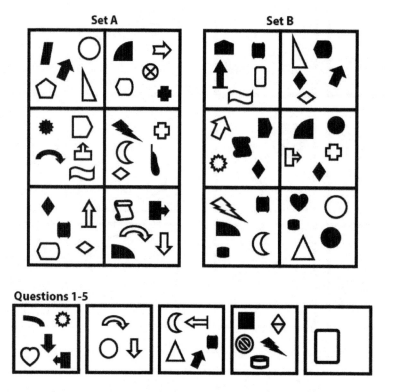

SET 2

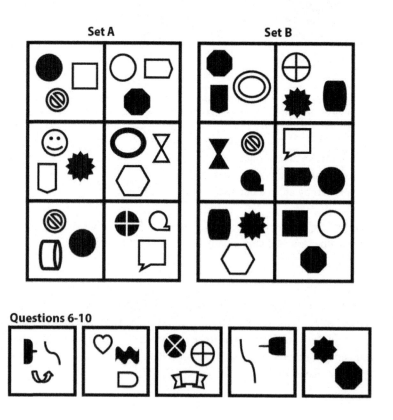

SET 3

SET 4

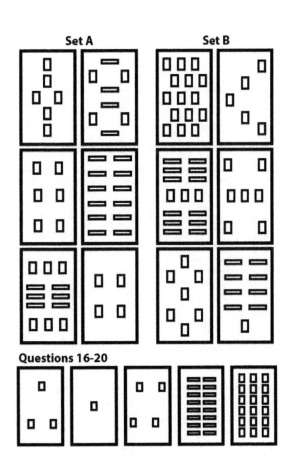

SET 5

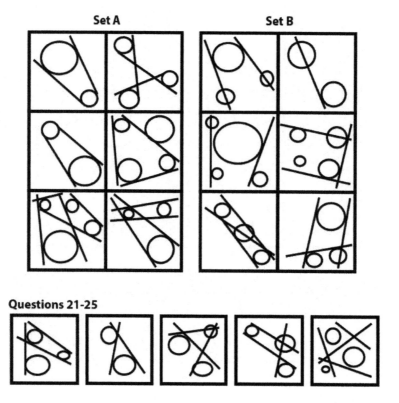

Questions 21-25

SET 6

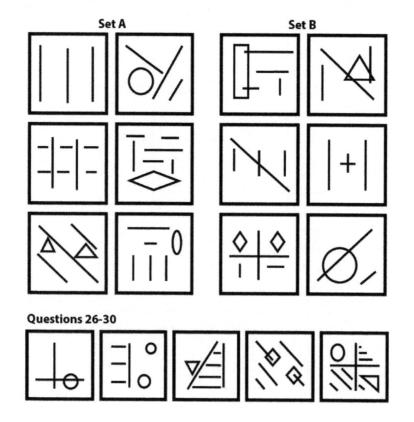

Questions 26-30

SET 7

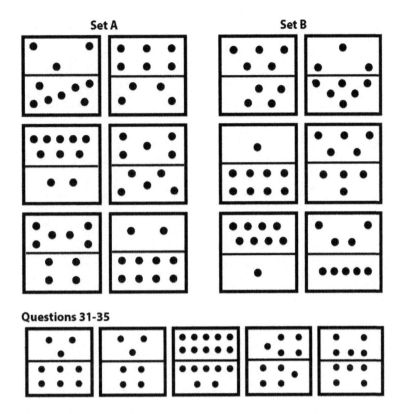

Questions 31-35

SET 8

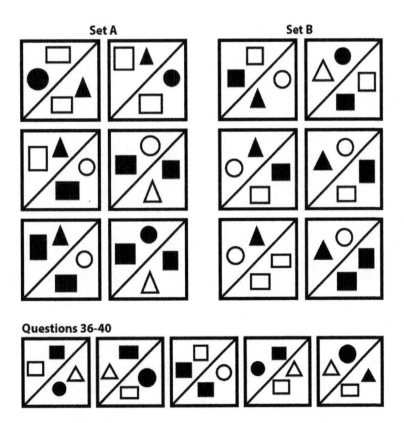

Questions 36-40

SET 9

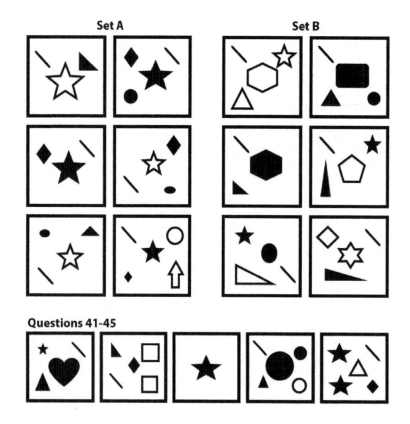

Set A   Set B

Questions 41-45

SET 10

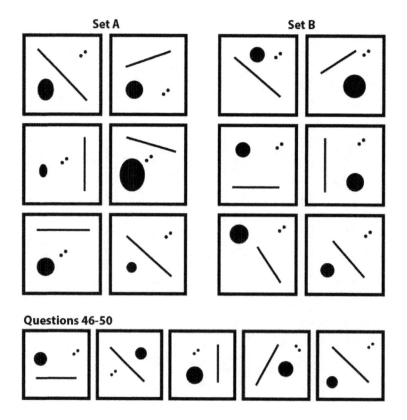

Set A   Set B

Questions 46-50

SET 11

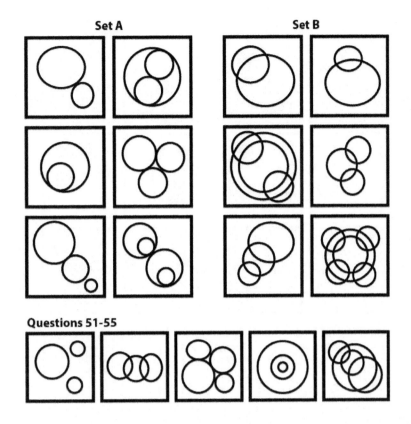

Questions 51-55

SET 12

Questions 56-60

SET 13

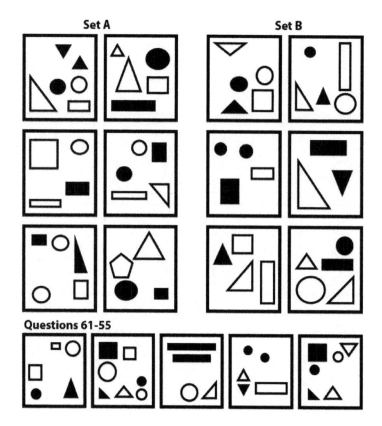

Questions 61-55

SET 14

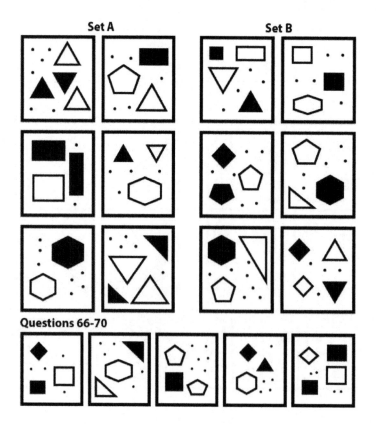

Questions 66-70

SET 15

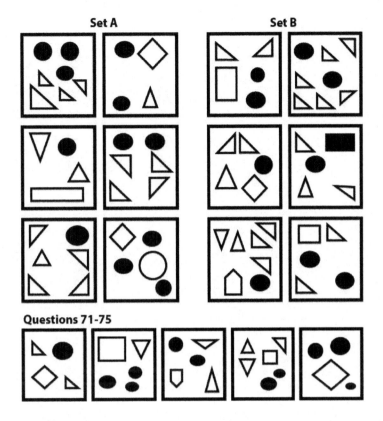

Questions 71-75

SET 16

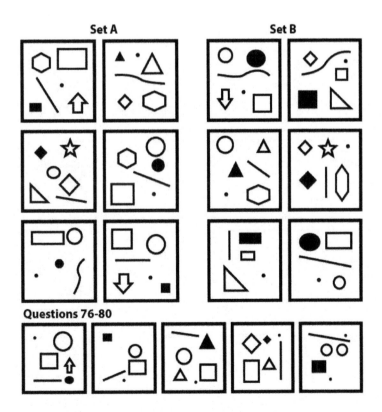

Questions 76-80

SET 17

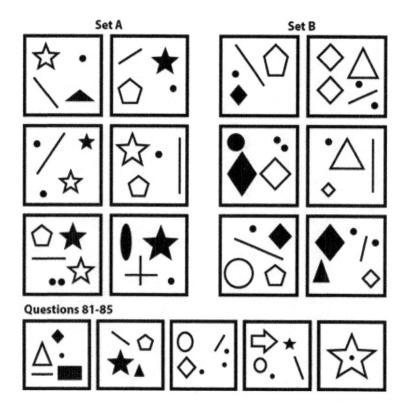

Questions 81-85

SET 18

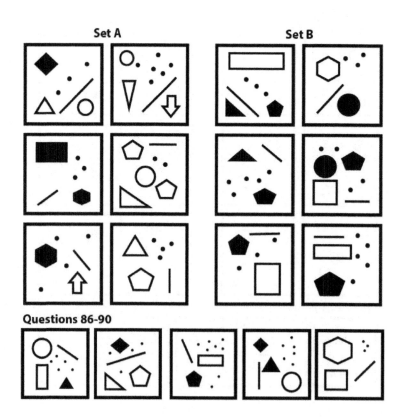

Questions 86-90

SET 19

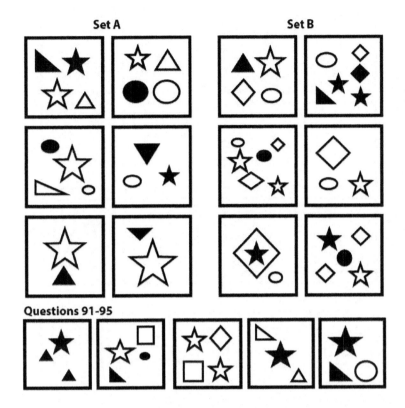

Questions 91-95

SET 20

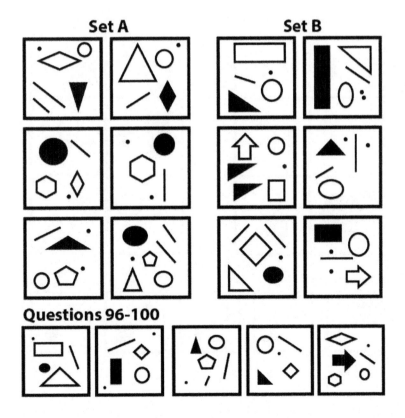

Questions 96-100

SET 21

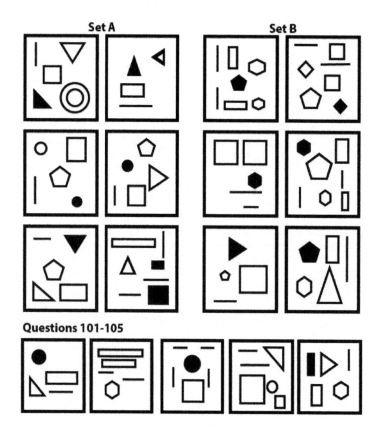

Questions 101-105

SET 22

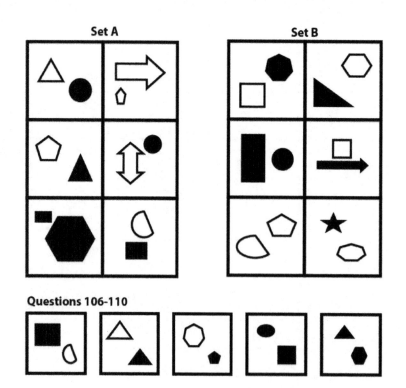

Questions 106-110

SET 23

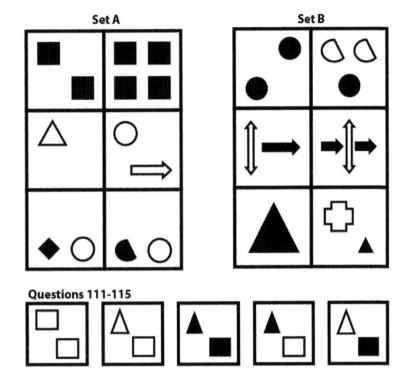

Questions 111-115

SET 24

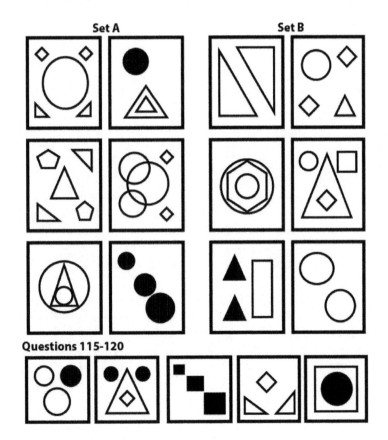

SET 25

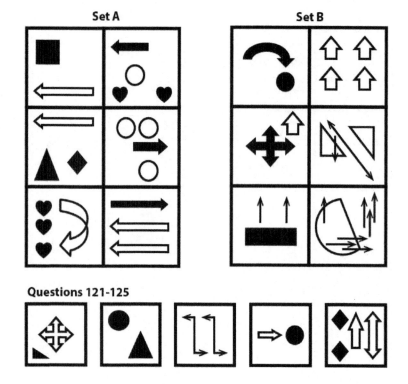

Questions 121-125

SET 26

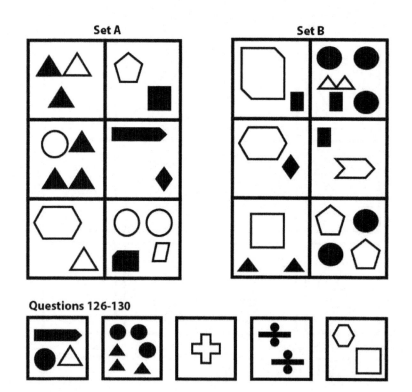

Questions 126-130

SET 27

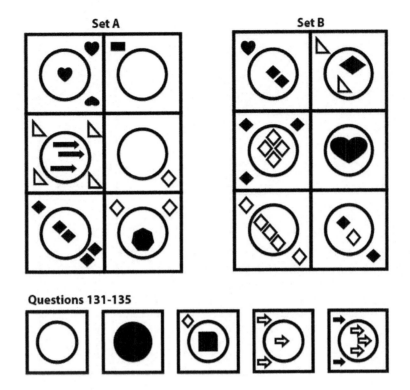

Questions 131-135

SET 28

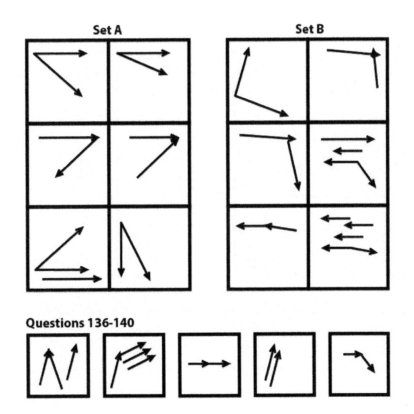

Questions 136-140

SET 29

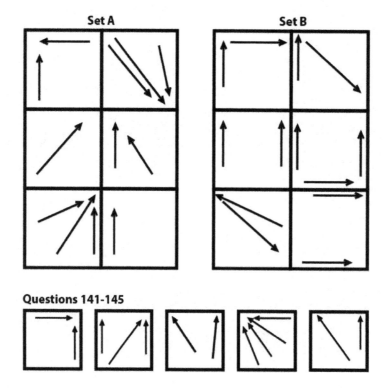

Questions 141-145

SET 30

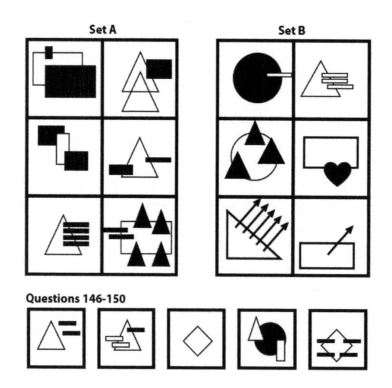

Questions 146-150

SET 31

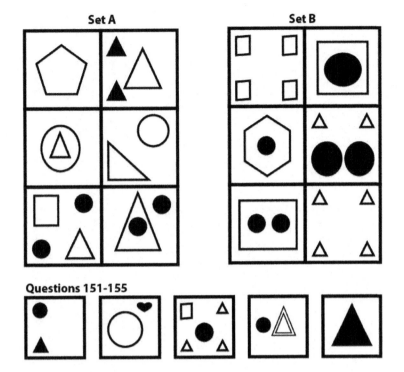

Questions 151-155

SET 32

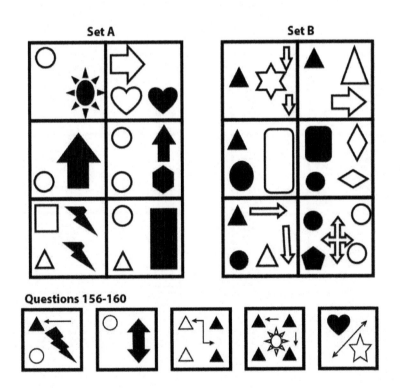

Questions 156-160

SET 33

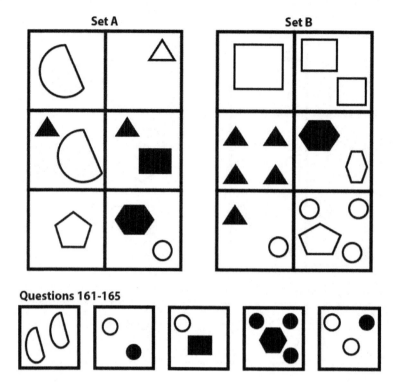

Questions 161-165

SET 34

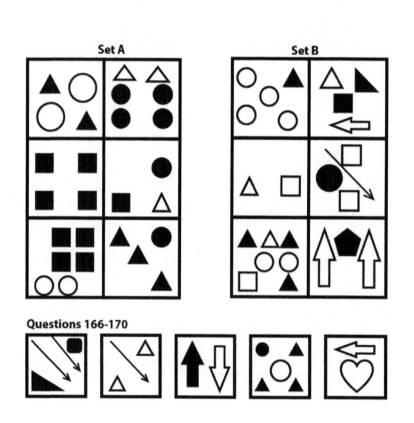

Questions 166-170

SET 35

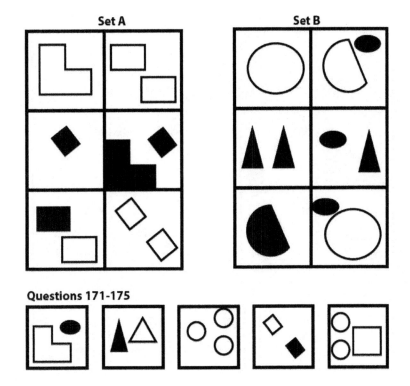

Questions 171-175

SET 36

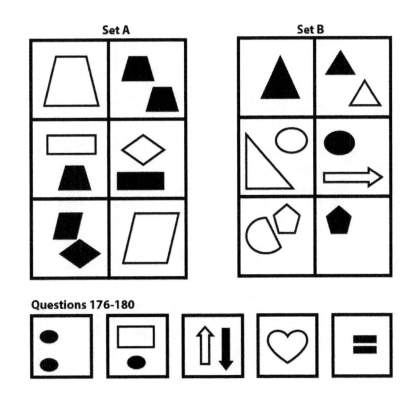

Questions 176-180

SET 37

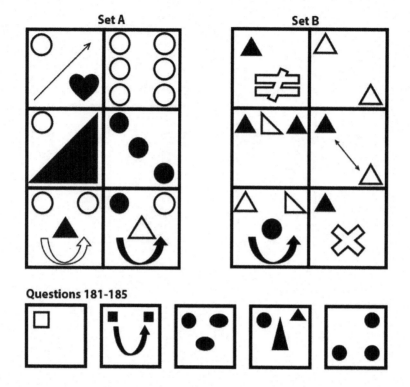

Questions 181-185

SET 38

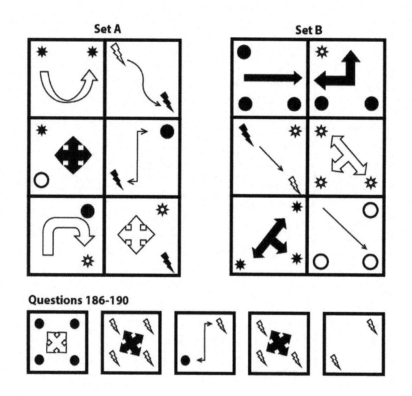

Questions 186-190

SET 39

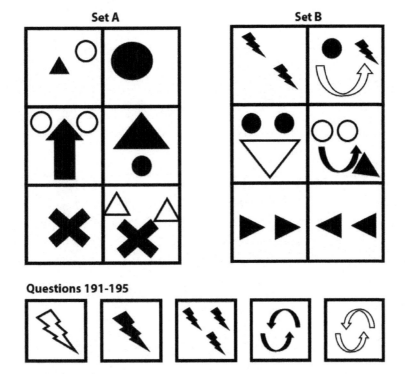

Questions 191-195

SET 40

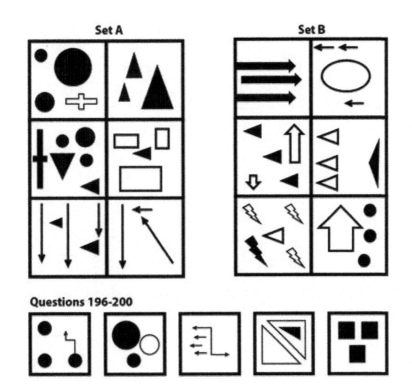

Questions 196-200

For Questions 201 – 250: which answer option completes the series?

Q201

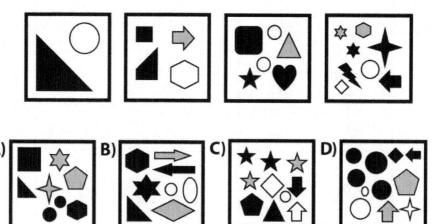

Q202

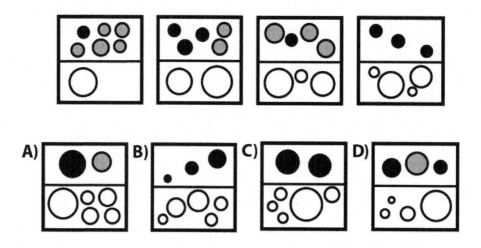

Q203

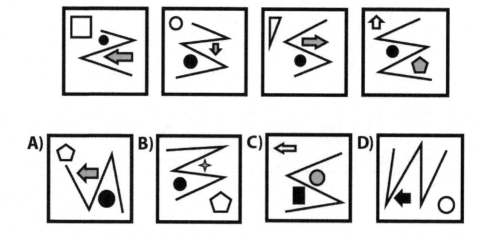

Q204

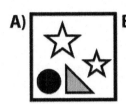 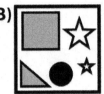 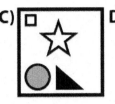 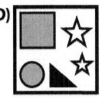

A) 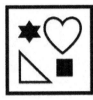 B) 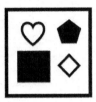 C) 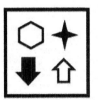 D)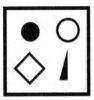

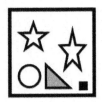 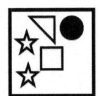 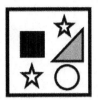

A) 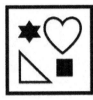 B) 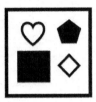 C) 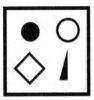 D)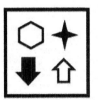

Q205

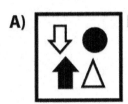 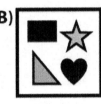 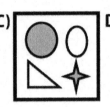 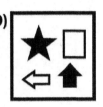

A) 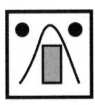 B) 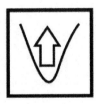 C) 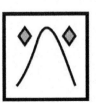 D)

Q206

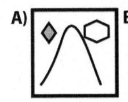 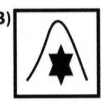 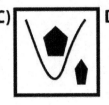 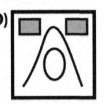

A) B) C) D)

Q207

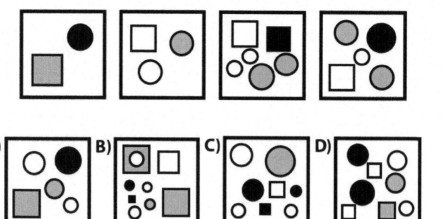

Q208

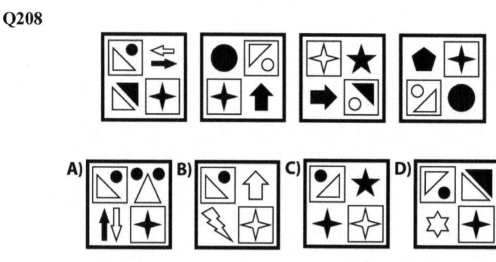

Q209

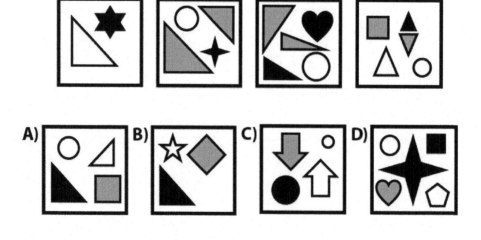

Q210

Q211

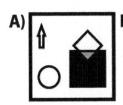 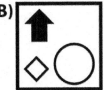 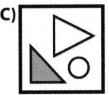 

   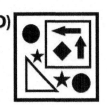

Q212

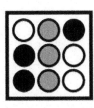 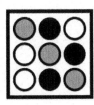 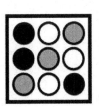 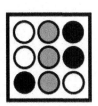

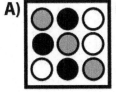 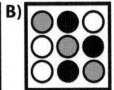 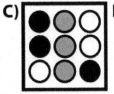 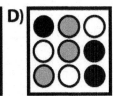

Q213

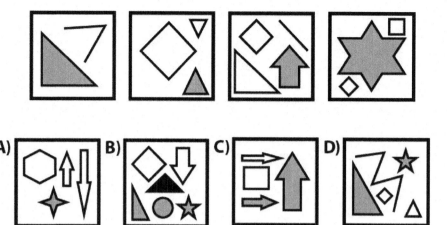

Q214

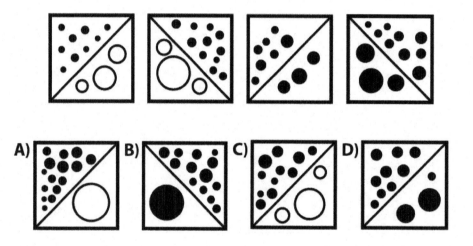

Q215

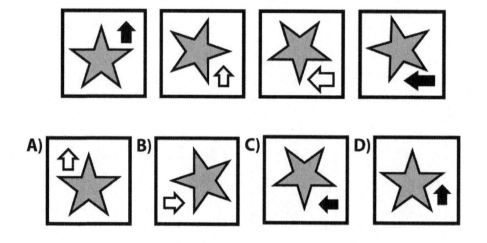

Q216

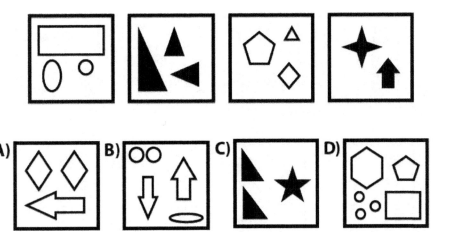

Q217

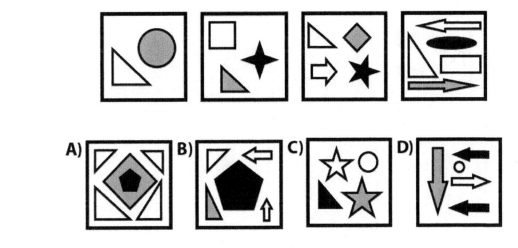

Q218

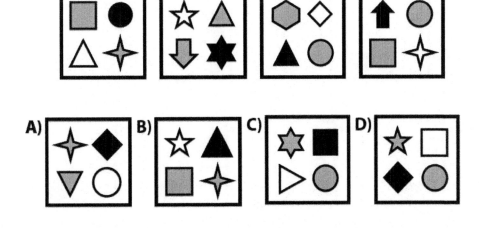

Q219

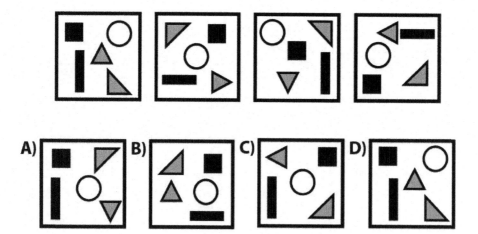

Q220

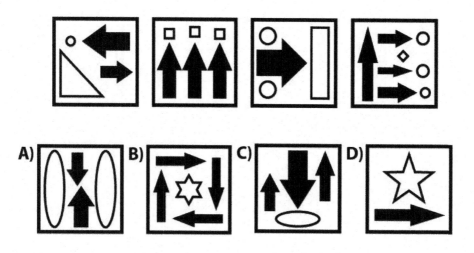

Q221

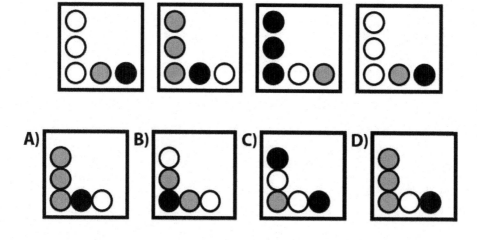

Q222

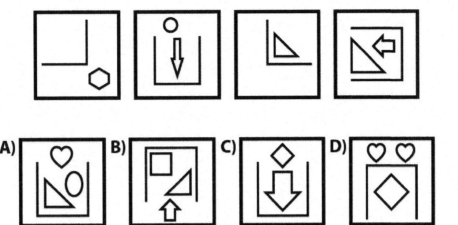

Q223

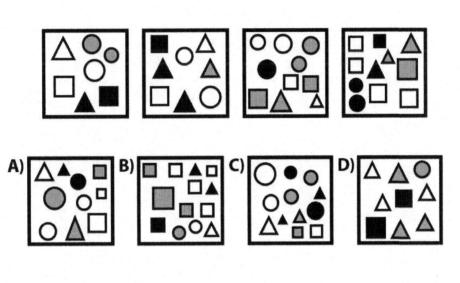

Q224

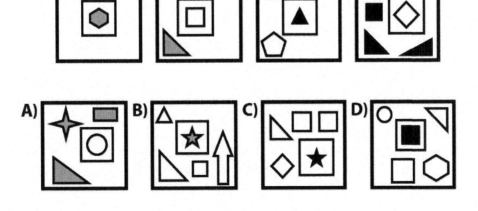

Q225

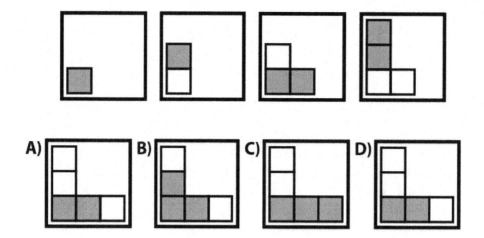

Q226

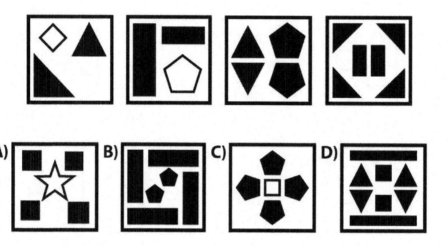

Q227

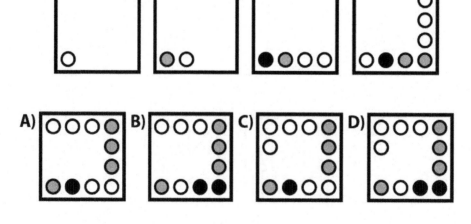

Q228

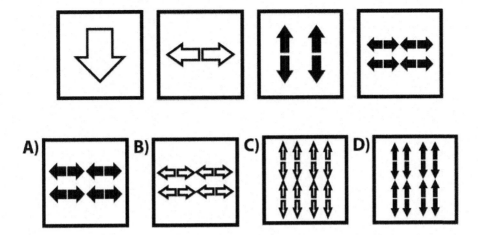

Q229

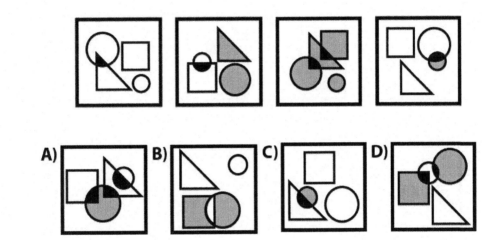

Q230

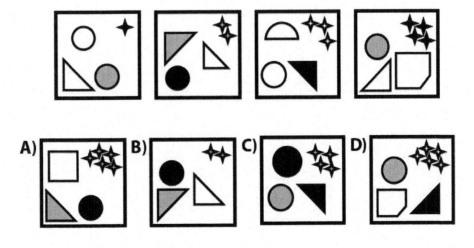

Q231

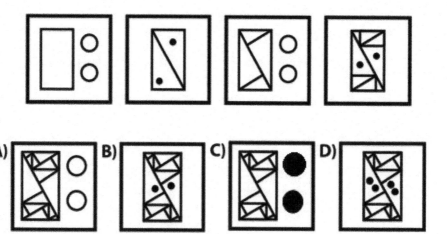

Q232

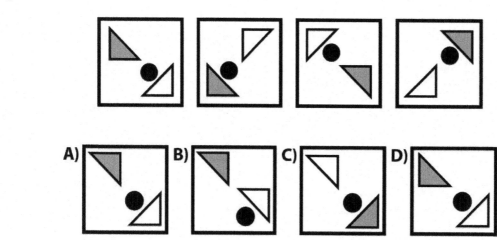

Q233

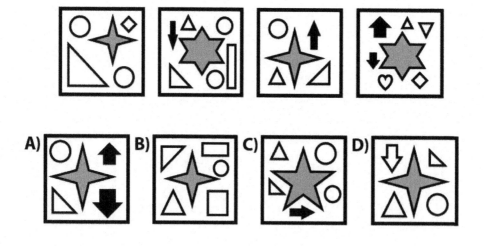

Q234

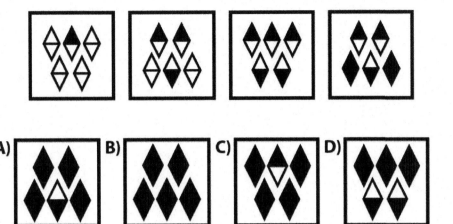

Q235

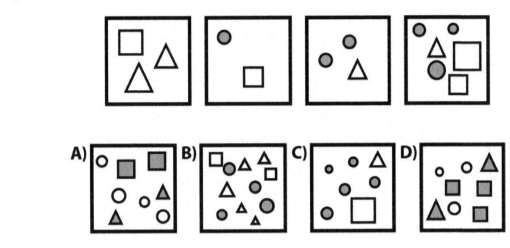

Q236

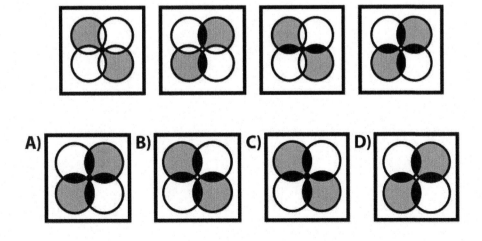

Q237

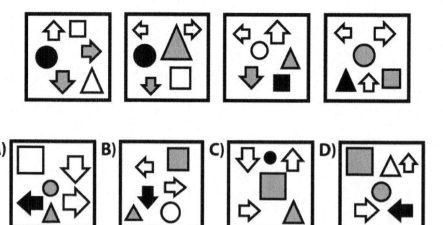

Q238

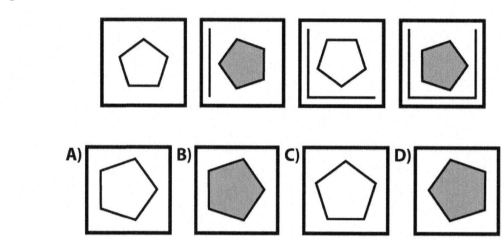

Q239

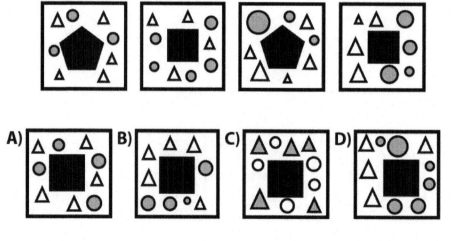

Q240

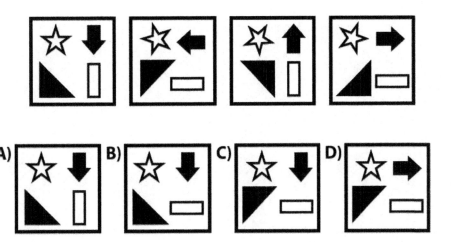

Q241

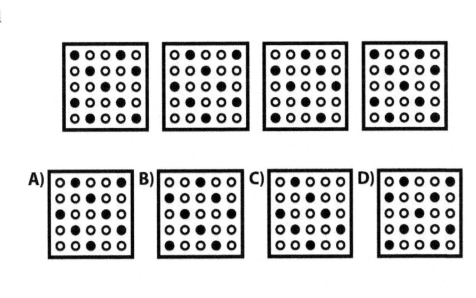

Q242

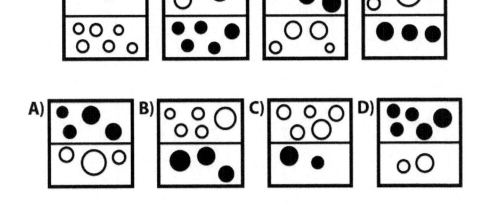

Q243

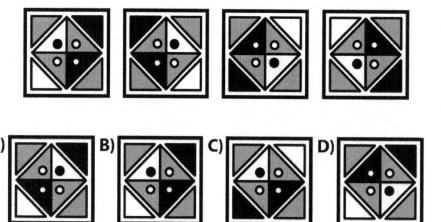

Q244

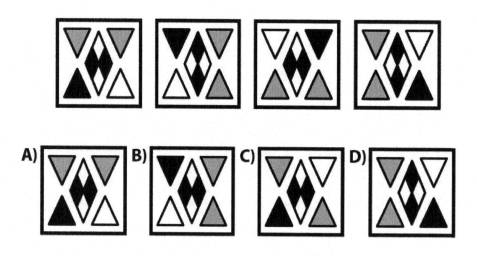

Q245

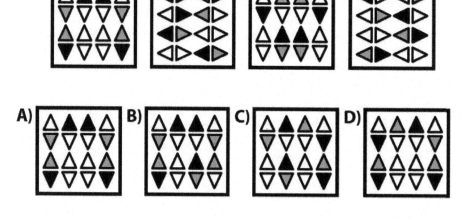

Q246

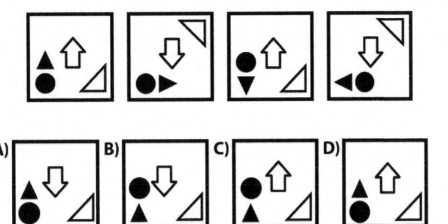

Q247

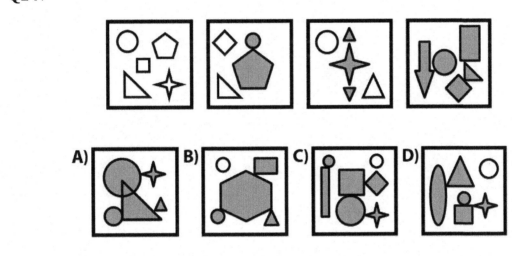

Q248

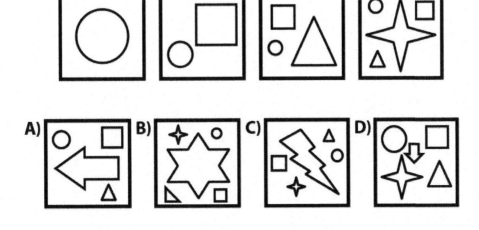

Q249

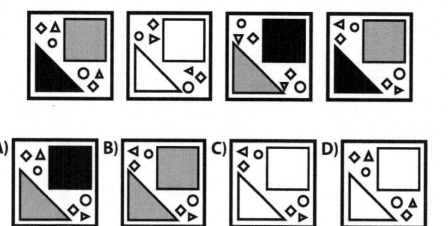

Q250

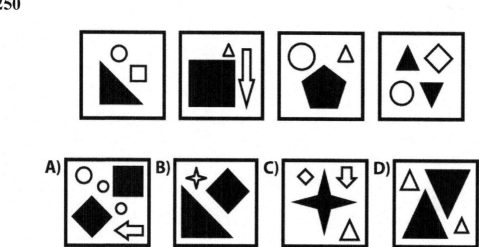

For Questions 251 – 300, which answer option completes the statement?

Q251

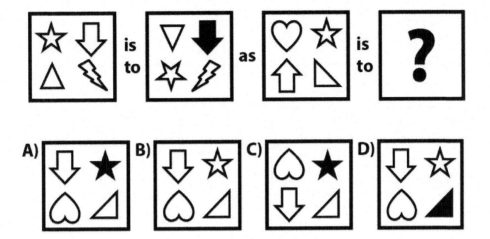

Q252

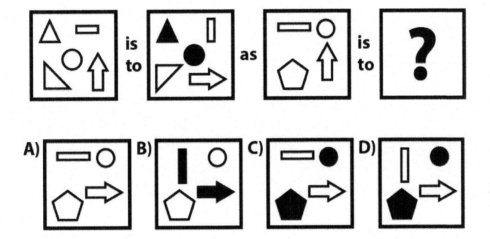

Q253

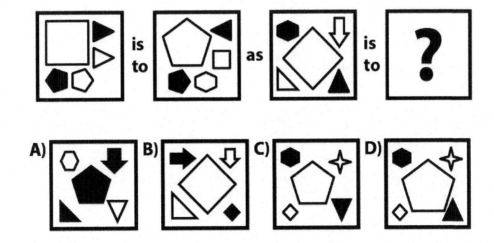

Q254

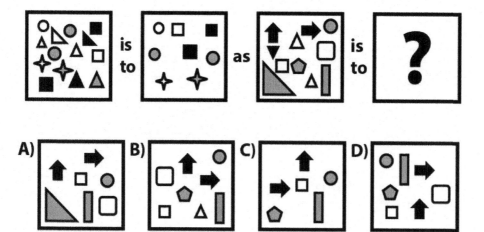

Q255

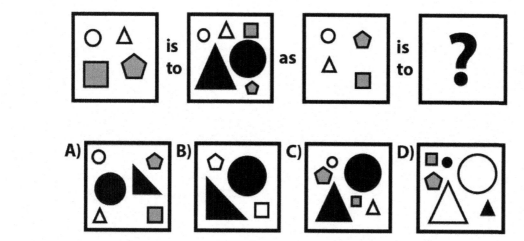

Q256

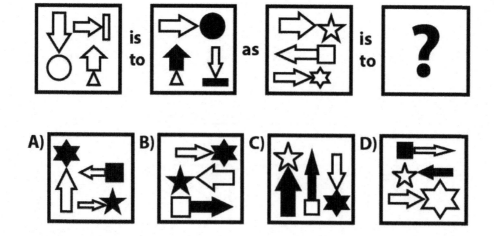

Q257

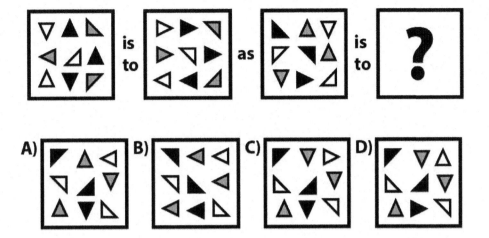

Q258

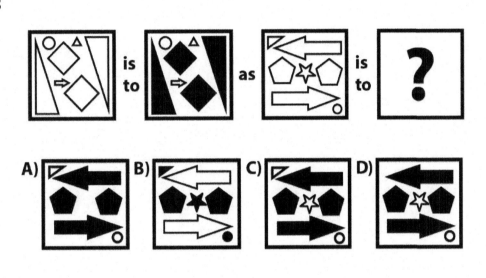

Q259

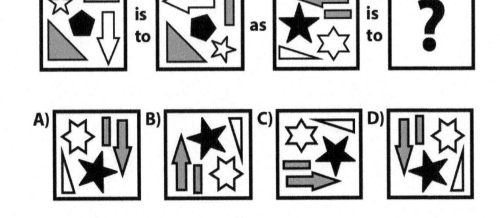

Q260

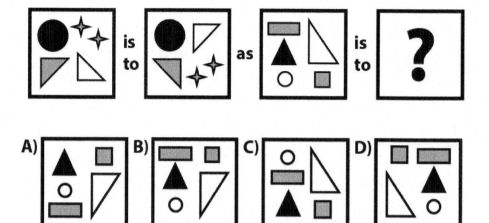

Q261

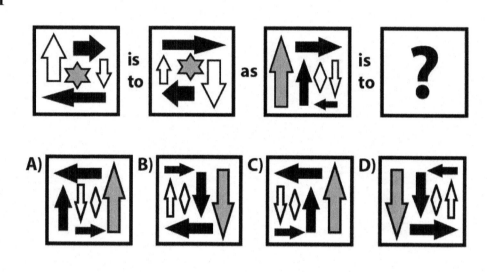

Q262

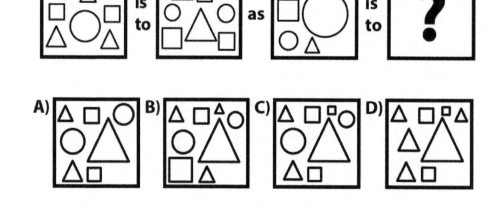

Q263

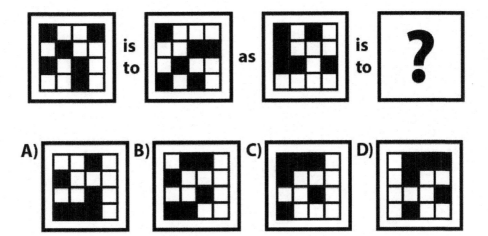

Q264

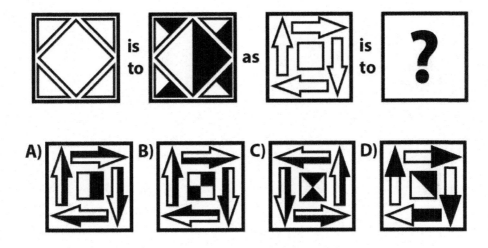

Q265

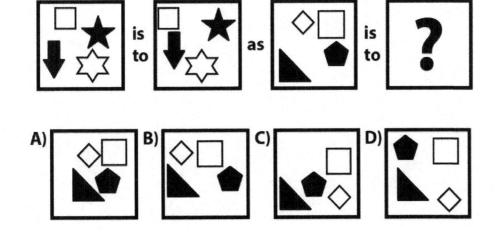

Q266

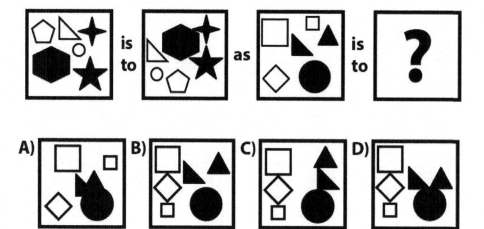

Q267

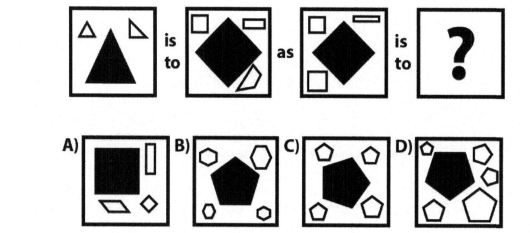

Q268

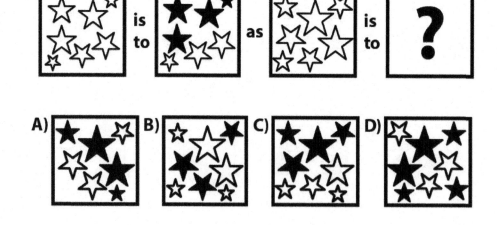

Q269

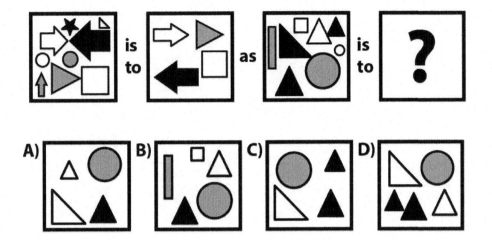

Q270

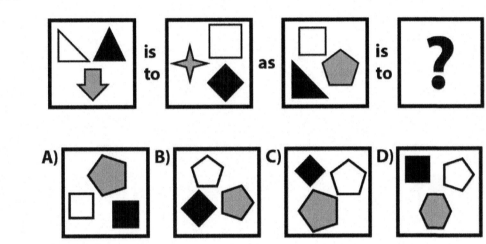

Q271

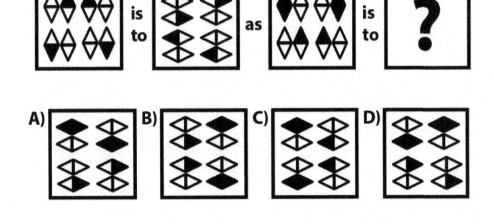

Q272

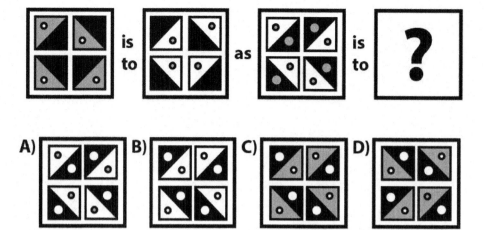

Q273

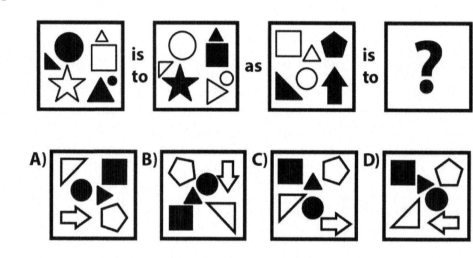

Q274

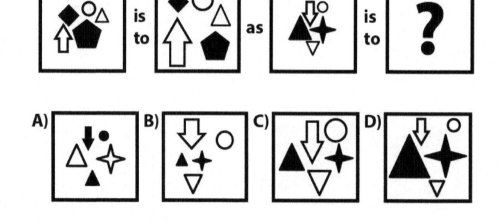

Q275

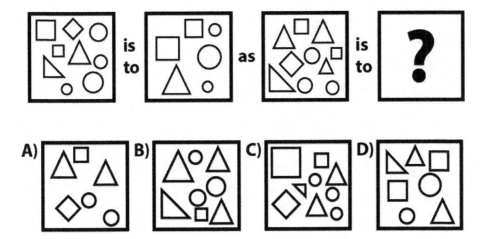

Q276

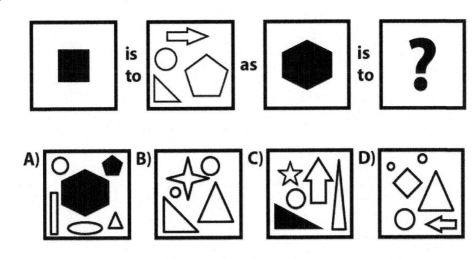

Q277

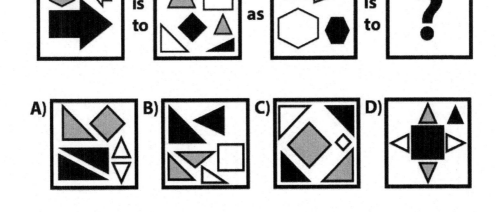

Q278

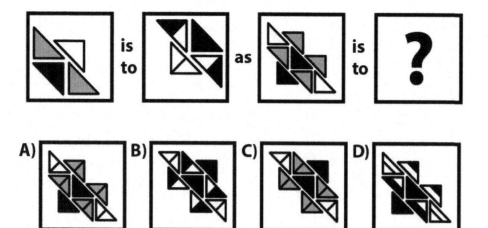

Q279

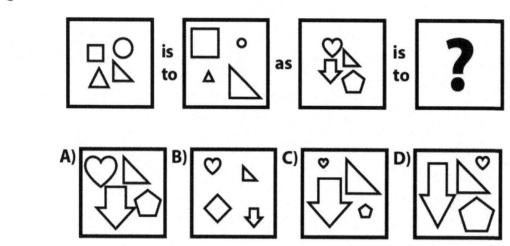

Q280

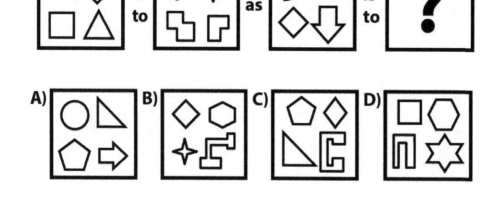

Q281

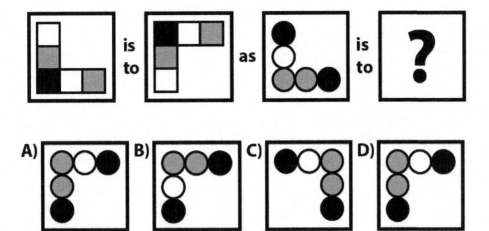

Q282

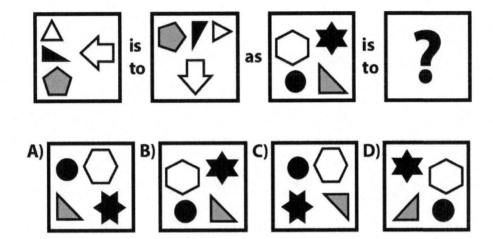

Q283

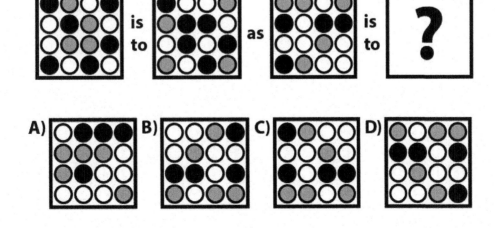

Q284

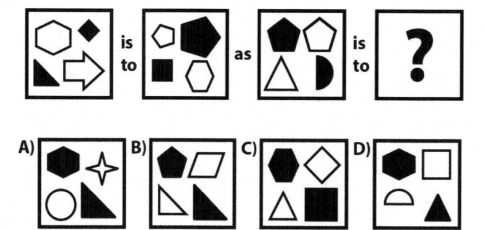

Q285

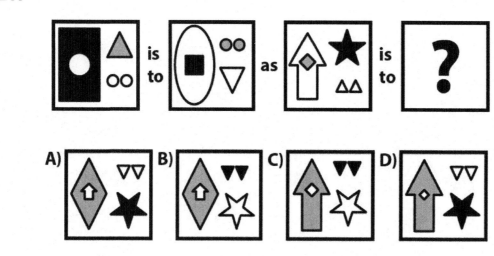

Q286

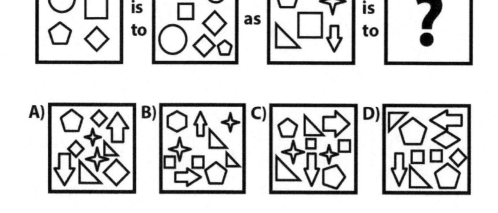

Q287

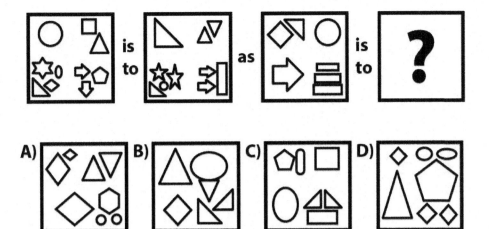

Q288

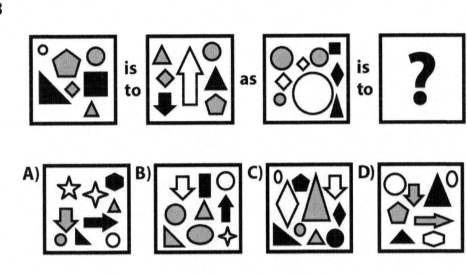

Q289

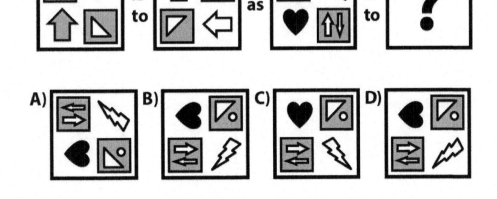

Q290

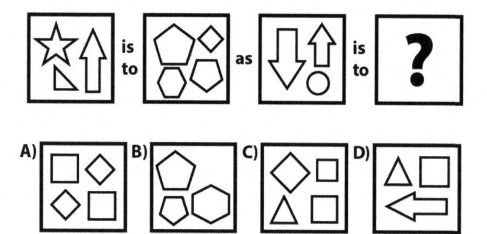

Q291

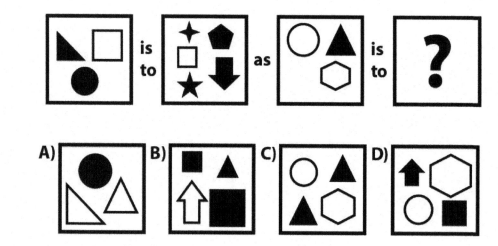

Q292

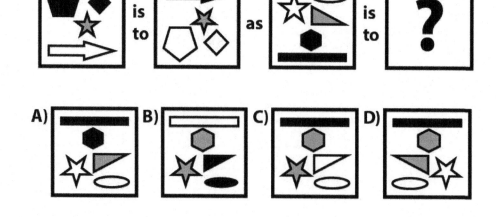

Q293

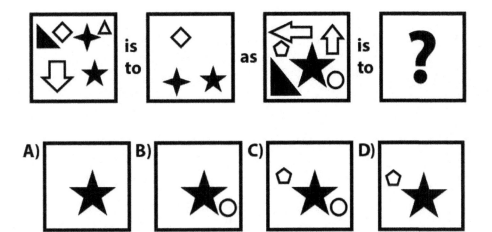

Q294

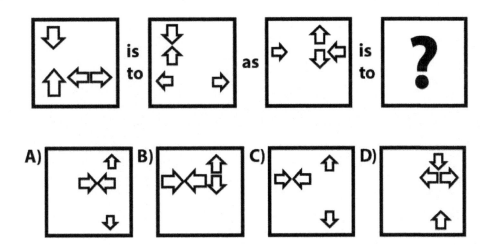

Q295

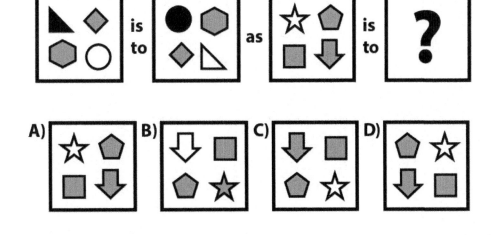

Q296

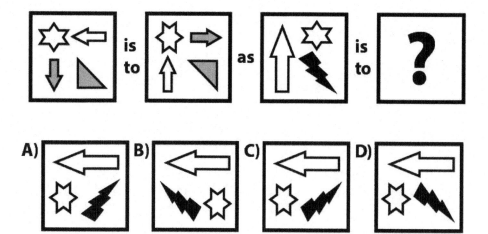

Q297

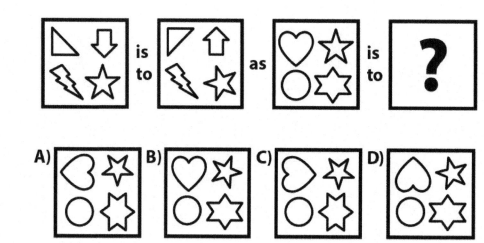

Q298

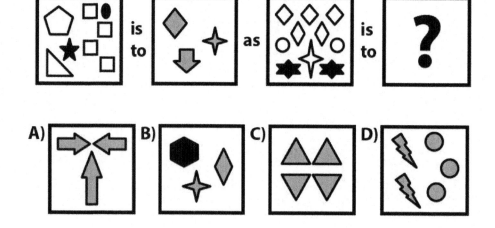

Q299

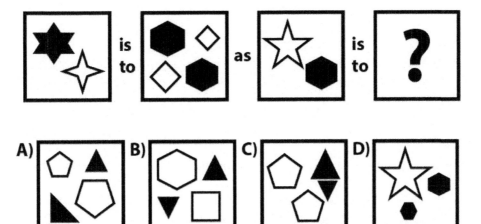

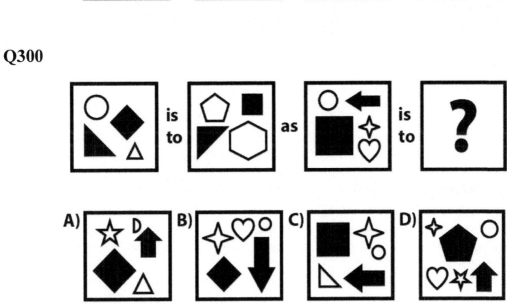

Q300

Situational Judgement

The Basics

Situational judgment is a psychological aptitude test; it is an assessment method used to evaluate your ability in solving problems in work-related situations. Situational Judgement Tests (or SJTs) are widely used in medicine as one of the criteria when deciding on applicants; it is used for the Foundation Programme and GP applications.

The aim of the situational judgment section is to assess your ability to understand situations that you could encounter as a medical student or doctor and how you would deal with them. It is a method to **test some of the qualities required in a healthcare professional** (e.g. integrity and ability to work in a team).

In the UKCAT, the situational judgment section consists of 20 scenarios with 67 items. Each scenario will have 3-6 items. You will have 26 Minutes to complete this section, which translates to approximately 22 seconds per item. As with all UKCAT sections, you have one additional minute to read the instructions at the start. Ensure you're careful to mark your intended answers when working at this pace.

This is the last section of your exam, you are almost near the end- only twenty seven more minutes to go. You still need to stay focused; it might seem obvious, but make sure you read the whole scenario and understand it prior to answering. When answering, imagine you are the person in the scenario. The majority of the scenarios will be about medical students, imagine you are in their shoes.

The series of scenarios include possible actions and considerations. Each scenario is comprised of two sets of questions. In set one you will be asked to assess the "appropriateness" of options in relation to the scenario.

The four possible <u>appropriateness</u> choices are:

➢ *A very appropriate thing to do* – This is an ideal action.
➢ *Appropriate, but not ideal* – This option can be done but not necessarily the best thing to be done.
➢ *Inappropriate, but not awful* – This should not be done, but if it does occur the consequences are not terrible.
➢ *A very inappropriate thing to do* – This should not be done in any circumstances, as it will make the situation worse.

In set two, you will be asked to assess the "importance" of options in relation to the scenario,

The four possible <u>importance</u> options are:

➢ *Very important* – something that is essential to take into account.
➢ *Important* – something you should take into account but is not vital.
➢ *Of minor importance* – something that may be considered, but will not affect the outcome if it is not taken into account.
➢ *Not important at all* – something that is not relevant at all.

Section E includes non-cognitive abilities and so is marked differently to the other sections. You will be awarded full marks if your response completely matches the correct answer. If your answer is close but not exactly right you will receive partial marks, and if there are no correct answers you will receive no marks. Your score is then calculated and expressed in one of four bands, band 1 being the highest and band 4 being the lowest. The band resembles how close your responses were to the assessment panel's agreed answers.

Things you MUST NOT do as a medical student:

➤ Write or sign drug charts.
➤ Sign or authorise anything that a doctor should do (operation consent forms, death certificates etc).
➤ Make any decisions that affect treatment.
➤ Change any treatment regime for any reason.
➤ Perform practical procedures without supervision.
➤ Break patient confidentiality or do anything that places confidentiality at risk.
➤ Behave dishonestly in any way.

Things you MUST do as a medical student:

➤ Raise any concerns regarding patient safety with an appropriate person.
➤ Report any inappropriate behaviour you witness to an appropriate person.
➤ Attend all scheduled teaching and training.
➤ Take responsibility for your learning and seek opportunities to learn.
➤ Dress and present yourself in a clean and smart way.

Things you CAN do as a medical student:

➤ Speak to patients.
➤ Examine patients.
➤ Write in patient notes (but must indicate your name and role).
➤ Perform simple practical procedures like taking blood, inserting cannulas or catheters (if you are trained to and with adequate supervision).
➤ Help doctors with more complex procedures according to their strict directions.
➤ Attend meetings where patients are discussed.
➤ Do things you have been trained to do.

> *Top tip!* Always put patient safety first. If you do this, you can never go far wrong.

Question Answering Strategy

Treat every option as independent – the options may seem similar, but don't let the different options confuse you, read each option as if it is a question on its own. It is important to know that responses should **NOT** be judged as though they are the **ONLY** thing you are to do. An answer should not be judged as inappropriate because it incomplete, but only if there is some actual inappropriate action taking place. For example, if a scenario says "a patient on the ward complains she is in pain", the response "ask the patient what is causing the problem" would be very appropriate, even though any good response would also include informing the nurses and doctors about what you had been told.

There might be multiple correct responses for each scenario, so don't feel you have to answer each stem differently. Thus an answer choice may be used once, more than once or not at all for all scenarios.

If you are unsure of the answer, mark the question and move on. Avoid spending longer than 30 seconds on any question, otherwise you will fall behind the pace and not finish the section.

As with every other section, if you are completely unsure of the answers, answer the question anyway. There is no negative marking and your initial instinct may be close to the intended answer.

➤ If there are several people mentioned in the scenario make sure you are answering about the correct person.
➤ Think of what you 'should' do rather than what you necessarily would do.
➤ Always think of **patient safety** and acting in the patient's best interests.

Read "Tomorrow's Doctors"

This is a publication produced by the GMC (General Medical Council) which can be found on their website. The GMC regulate the medical profession, ensuring standards remain high. This publication can be found on their website, and it outlines the expectations of the next generation of doctors – the generation of doctors you are aspiring to join. **Reading through this will get you into a professional way of thinking** that will help you judge these questions accurately.

Step into Character

When doing this section, imagine you're there. Imagine yourself as a caring and conscientious medical student a few years from now, in each situation as it unfolds. What would you do? What do you think would be the right thing to do?

Hierarchy

The patient is of primary importance. All decisions that affect patient care should be made to benefit the patient. Of secondary importance are your work colleagues. So if there is no risk to patients, you should help out your colleagues and avoid doing anything that would undermine them or harm their reputation – but if doing so would bring detriment to any patient then the patients priorities come to the top. Finally of lowest importance is yourself. You should avoid working outside hours and strive to further your education, but not at the expense of more important or urgent priorities. Remember the key principles of professional conduct and you cannot go far wrong. **Of first and foremost importance is patient safety**. Make sure you make all judgements with this in mind.

> *Top tip!* Read the GMC publication "*Tomorrow's Doctors*" – this will help you think the right way.

Medical Ethics

There tend to be a few ethical questions in each SJT paper so it is well worth your time to learn medical ethics. Whilst there are huge ethical textbooks available– you only need to be familiar with the basic principles for the purposes of the UKCAT. These principles can be applied to all cases regardless what the social/ethnic background the healthcare professional or patient is from. In addition to being helpful in the UKCAT, you'll need to know them for the interview stages anyway so they're well worth learning now rather than later. The principles are:

Beneficence

The wellbeing of the patient should be the doctor's first priority. In medicine this means that one must act in the patient's best interests to ensure the best outcome is achieved for them i.e. 'Do Good'.

Non-Maleficence

This is the principle of avoiding harm to the patient (i.e. Do no harm). There can be a danger that in a willingness to treat, doctors can sometimes cause more harm to the patient than good. This can especially be the case with major interventions, such as chemotherapy or surgery. Where a course of action has both potential harms and potential benefits, non-maleficence must be balanced against beneficence.

Autonomy

The patient has the right to determine their own health care. This therefore requires the doctor to be a good communicator, so that the patient is sufficiently informed to make their own decisions. 'Informed consent' is thus a vital precursor to any treatment. A doctor must respect a patient's refusal for treatment even if they think it is not the correct choice. Note that patients cannot <u>demand</u> treatment – only refuse it, e.g. an alcoholic patient can refuse rehabilitation but cannot demand a liver transplant.

There are many situations where the application of autonomy can be quite complex, for example:

> **Treating Children**: Consent is required from the parents, although the autonomy of the child is taken into account increasingly as they get older.

> **Treating adults without the capacity** to make important decisions. The first challenge with this is in assessing whether or not a patient has the capacity to make the decisions. Just because a patient has a mental illness does not necessarily mean that they lack the capacity to make decisions about their health care. Where patients do lack capacity, the power to make decisions is transferred to the next of kin (or Legal Power of Attorney, if one has been set up).

Justice
This principle deals with the fair distribution and allocation of healthcare resources for the population.

Consent
This is an extension of Autonomy- patients must agree to a procedure or intervention. For consent to be valid, it must be **voluntary informed consent.** This means that the patient must have sufficient mental capacity to make the decision, they must be presented with all the relevant information (benefits, side effects and the likely complications) in a way they can understand and they must make the choice freely without being put under pressure.

Top tip! Remember that **consent is only valid** if it is given:

> On the basis of full information.
> With sufficient mental capacity.
> Freely without pressure.
> Communicated unambiguously.

Confidentiality
Patients expect that the information they reveal to doctors will be kept private- this is a key component in maintaining the trust between patients and doctors. You must ensure that patient details are kept confidential. Confidentiality can be broken if you suspect that a patient is a risk to themselves or to others e.g. Terrorism, suicides.

When answering a question on medical ethics, you need to ensure that you show an appreciation for the fact that there are often two sides to the argument. Where appropriate, you should outline both points of view and how they pertain to the main principles of medical ethics and then come to a reasoned judgement.

Situational Judgement Questions

Scenario 1

A conversation is taking place between a midwife Kate and the senior Dr Herbert: Jacob, the medical student, is observing. Dr Herbert is being rude to the Kate and is acting superior. When Dr Herbert leaves, Jacob overhears Kate talking to the other midwives about his behaviour, and how it happens frequently, and makes both the midwives and the patients feel uncomfortable.

How appropriate are the following actions from Jacob?

1. Tell Kate that you will help to file a complaint against Dr Herbert.
2. Make Dr Herbert aware that perhaps he should be kinder the next time he speaks to Kate and patients.
3. Ignore the situation and do nothing.
4. Alert his supervisor as to what he saw, and to get advice on what to do.
5. Tell Dr Herbert that his behaviour was making patients and midwives feel uncomfortable.

Scenario 2

A medical student, George, is sitting in a foot clinic with Dr Walker. George notices that Dr Walker hasn't been washing his hands between patients, despite examining the feet of all of his patients without gloves. In his training George was told that he must wash his hands properly before and after touching each patient to prevent the spread of infections.

How appropriate are each of the following responses by George in this situation?

6. Alert Dr Walker that he ought to wash his hands more after the current consultation has finished.
7. Wash his hands before and after each patient in the hopes that Dr Walker will follow by example.
8. Do nothing because Dr Walker is an experienced consultant.
9. Tell the nurse in charge of the foot patients after the clinic has finished.
10. Write in the patient notes that Dr Walker didn't wash his hands before examining them.

Scenario 3

A medical student, Linh, is working on a project with a small group of other students. The students have to examine real skull bones, which were provided by the medical school's museum, and are very valuable. One of the students in Linh's group accidentally drops the skull and some of the smaller delicate bones shatter.

How appropriate are the following responses by Linh?

11. Ignore what happened, throw the skull remains away, and borrow another group's skull to finish the project.
12. Alert the museum curator about what happened as a group, and write a letter of apology together.
13. Pretend that the skull was stolen.
14. Tell the museum curator in private about who dropped the skull.
15. Tell her supervisor.

Scenario 4

A medical student, Henry, is living in a set of halls with students that study many different subjects. The other students find it funny to joke about Henry's work. Henry is finding it difficult to keep up with his work, and silently takes offense every time the other students joke with him. The night before one of Henry's exams, the other students make a joke that really affects Henry, and he is unable to concentrate on finishing up his revision.

How <u>appropriate</u> are each of the following responses by <u>Henry</u>?

16. Speak to his personal tutor about how he can organise himself and tackle his work in the future.
17. Retaliate by insulting the other students.
18. Do nothing because he doesn't want to offend anyone and is embarrassed about not being able to cope with the workload.
19. Move out of the halls.
20. Speak to his medical student friends about how annoying he finds his flat mates.

Scenario 5

Mark, a medical student, is working with a group of nursing and physiotherapy students to learn about integrated care. Mark is mistaken for a junior doctor, as he is not in uniform, and is asked to test the urine of an elderly patient on the ward using a dipstick. Mark is familiar with the patient, and knows exactly how to do the test. Unfortunately, the doctor that asked him to do the test has left, and there are no other members of staff that are able to do the test for another 5 hours. The results of the test will determine the patient's management.

How <u>appropriate</u> are the following responses by <u>Mark</u>?

21. Get the most senior student in his study group to perform the test and write the results in the patient's notes.
22. Do the test himself and write the results in the patient's notes.
23. Bleep the doctor that is in charge of the patient to alert him about his mistake.
24. Pretend that the doctor never asked him to do the test.
25. Try to find another member of staff that would be capable of performing the test.

Scenario 6

A medical student, Adele, is studying for her first year exams. She has started to panic, and does not feel as though she will be able to complete her revision before the exams start. If Adele fails the exams she would have to resit them in her holidays, which she has come to terms with. She is embarrassed of the possibility of failing, and would rather tell her friends and family that she was ill and unable to take the exams than face the embarrassment of failure. It is against the Medical School rules to opt out of an exam without a medical reason and a Doctor's letter.

How <u>appropriate</u> are the following actions for <u>Adele</u> to take?

26. Fake an illness and postpone her exams.
27. Speak to her parents and her personal tutor about her struggle to get through the revision.
28. Speak to the other medical students to see if they all felt the same way about their work.
29. Refuse to turn up to the exams on the day and pretend that she had food poisoning.
30. Make an efficient revision plan for her remaining days before the exams and attempt to do the exams as best as she can.

Scenario 7

Rohan, a final year medical student, notices that Dijam, one of the medical students on his ward who had been drinking a lot the previous night is on call.

How <u>appropriate</u> are the following actions by <u>Rohan</u>?

31. Advise Dijam to go home.
32. Ignore the situation because Dijam wasn't actually treating any of the patients.
33. Inform the doctors that are on call with Dijam.
34. Joke with Dijam about how he managed to make it into work on time.
35. Inform the Doctor that is in charge of Dijam and Rohan's attachment about Dijam's state.

Scenario 8

Patrick is a medical student, and is working with another group of students on a project that they will receive a joint mark for. Patrick has noticed that there are a couple of loud dominating people in the group, and that the rest of the group are very shy and quiet, and rarely contribute to the conversations. Jina is one particularly loud student that is involved, however she has been making some excellent points and is happy to do a lot of the work.

How <u>appropriate</u> are the following responses by <u>Patrick</u>?

36. Ignore the situation and allow Jina to do the majority of the work.
37. Ask his personal tutor for advice on how he should tackle the situation.
38. Ask the quieter members of the team about what they think of Jina.
39. Subtly hint to everyone to try to contribute more during the sessions so that it is a more even contribution from everyone.
40. Confront Jina and tell her to be less dominating during the sessions.

Scenario 9

Nazia, a medical student, has been working on busy hospital ward. She has been writing up notes from the patient's notes into her notebook so that she can construct a presentation on the case for her study group. No-one is allowed to remove the patient's notes from the hospital. However, she has noticed that one of her friends, Joshua, has a set of patient's notes sticking out of his bag. He has an appointment to get to, and has no time to write the notes up whilst at the hospital. Joshua says that he will return the notes first thing in the morning after he completes the work at home.

How <u>appropriate</u> are the following responses from <u>Nazia</u>?

41. Tell Joshua to do the presentation the next day instead when he has more time.
42. Ignore what he is doing.
43. Tell the ward nurses after Joshua leaves with the notes.
44. Seek advice from your clinical supervisor.
45. Tell Joshua that you will write the notes for him so he doesn't have to take the notes away from the hospital.

Scenario 10

Mr. Marshall has been seeing Dr Kelly regularly for years to check up on his diabetes. Recently, Mr Marshall has been seen by a different specialist doctor, Dr O'Brien. Dr O'Brien runs a test that shows that Mr. Marshall has cancer. He is then booked to see Dr O'Brien the following week who will break the diagnosis. Mr. Marshall is currently in clinic with Dr Kelly and asks her "is everything okay?"

How <u>appropriate</u> are the following responses by <u>Dr Kelly</u>?

46. Tell Mr. Marshall that everything is fine.
47. Reassure Mr. Marshall that Dr O'Brien will be able to answer his question better.
48. Tell Mr. Marshall that she is not allowed to discuss that information with him yet.
49. Look away and say nothing to try to express the seriousness of the situation.
50. Tell Mr. Marshall that he has cancer.

Scenario 11

Mary, a patient, has been in hospital for a long time whilst she recovers from a leg wound, and is desperate to return home. One day, Dr Anil is speaking to her on the ward. He has to leave urgently to answer his bleep call. Mary is left with a junior medical student, Julia. She asks Julia why she is still in the hospital, and wants to know if she can leave that day.

How <u>appropriate</u> are the following responses by <u>Julia</u>?

51. Explain to Mary that she is unable to answer her question, and that the doctor will be back soon.
52. Tell Mary that in most cases people wouldn't be able to leave the hospital at her stage of recovery.
53. Tell Mary that she can self-discharge from hospital if she is very keen to leave, but that it might be against medical opinion.
54. Tell Mary that she will find out and let her know.
55. Answer Mary's questions directly.

Scenario 12

Daniel, a first year medical student, is visiting a hospital for the first time since he started medical school. The doctor supervising them asked Hannah, another student to let the group know that they would be having a hand washing assessment consisting of practical and theoretical aspects. Unfortunately, Daniel was not been about the assessment, and doesn't know how to wash his hands properly.

How <u>appropriate</u> are the following responses by <u>Daniel</u>?

56. Ask to have his turn once a few of his colleagues had been so he can observe.
57. Confront Hannah and ask her why she didn't tell him about the assessment.
58. Tell the doctor that it was Hannah's fault.
59. Ask the group to see how many people were prepared for the assessment.
60. Ask the doctor if he can have his assessment another day so he can learn the skill and the theory properly.

Scenario 13

Helen, a medical student, is waiting for her exam results. She was very worried because she didn't feel as though she was ready for them. When the results come out, she realises that she has to retake her exams. She had booked to go travelling with a friend in South America over the summer holidays, but the resit exams are during the holidays and she is now worried that she will be unable to go, or that if she goes she will not have enough time to revise for the exams. She is also worried that her friend will be left to travel by herself if she doesn't go.

How <u>appropriate</u> are the following responses by <u>Helen</u> in this situation?

61. Call her friend and cancel the holiday.
62. Go travelling but take her revision with her and revise every day.
63. Go for part of the holiday and come home early to revise for the exams.
64. Go travelling and cram in the revision in the few days between coming back and taking the exams.
65. Try to get another friend to replace her so she can stay at home and revise but without leaving her friend to fend for herself.

Scenario 14

Celia, a medical student, is living at home instead of at halls because she doesn't live far away from the medical school. She found it hard to make friends in first year, and wants to move out for her second year or she fears that she will be further isolated from everyone. Unfortunately, that would depend on financial support from her parents. Celia's parents are unable to provide much financial aid, and Celia doesn't have time to take up another job.

How <u>appropriate</u> are the following responses by <u>Celia</u>?

66. Confront her parents and demand the money because they are 'denying her a student experience'.
67. Live at home but join a sports team so she can meet more people and join in with the student lifestyle a little more.
68. Start spending every night at her friend's room in halls.
69. Live at home in resentment and isolate herself from her university friends.
70. Come to an agreement with her parents that if she can move out for a couple of years and then live at home for the rest of medical school.

Scenario 15

Xun, a medical student, is due to hand in an essay the following day at 8AM but is only half way to finishing it at 9PM. The essay will contribute 20% to his final grade and he is beginning to panic.

How <u>appropriate</u> are the following responses by <u>Xun</u>?

71. Stay up late and finish the essay so that he doesn't miss the deadline.
72. Call the head of the assessments and explain his situation to them, in the hope that there will be some leniency.
73. Give up and hand in the essay half-complete.
74. Copy out a similar essay that a friend has written.
75. Fake an illness and ask for an extension.

Scenario 16

Nahor, a medical student, has always enjoyed having creative hairstyles. He is starting his rotations in the hospitals next week, and is worried that he will be unable to express himself through his hair anymore.

How <u>appropriate</u> are the following responses by <u>Nahor</u> in this situation?

76. Cut his hair and get a professional looking colour and style.
77. Start his hospital rotations with his pink long hair, and refuse to change it if he is asked to.
78. Start his hospital rotations with his pink long hair, and only change it if he is asked to.
79. Keep his hair a little quirky but make it look more professional that it has previously been.
80. Request permission from the clinical dean to keep his hair as it is.

Scenario 17

Charles, a medical student, was on call with Dr Patel in a busy hospital. Dr Patel told Charles to wait for the doctor that is going to handover to arrive before he leaves. The doctor isn't due to arrive for another 30 minutes. Charles has a sports match in 15 minutes, and needs to leave before then. Unfortunately Dr Patel is busy with a patient and is not answering his bleep.

How <u>appropriate</u> are the following responses by <u>Charles</u>?

81. Leave and email Dr Patel with a letter of apology.
82. Call up the captain of the sports team to apologise, and promise that you will make the second half of the match.
83. Leave the on call room and try to find Dr Patel to talk to him in person.
84. Leave a message with one of the nurses in the on call room to tell the doctor that is meant to be taking over.
85. Try to call up the doctor that is taking over to explain the situation to him.

Scenario 18

Archie, a medical student, is in clinic with Dr Coombe. Dr Coombe explains to the patient that her medication isn't working and that will have to try something else. Dr Coombe has to take an urgent call and walks out of the room- leaving Archie and the patient in the room. The patient then starts to ask Archie lots of questions about her medication.

How <u>appropriate</u> are the following responses by <u>Archie</u>?

86. Explain that he is unable to say, but that the patient should direct her questions towards Dr Coombe when he returns.
87. Try to answer the questions to the best of his ability.
88. Tell the patient that they should 'Google' the answers.
89. Tell the patient that he will 'Google' the answers.
90. Excuse himself and walk out of the room to leave the patient in there by herself until Dr Coombe returns.

Scenario 19

Matthias, a medical student, has hurt his knee whilst playing hockey. He will need to wear a full leg brace which will prevent him from walking around the hospital. Matthias is concerned that it will affect his studies adversely. He will have to take at least 6 weeks off.

How <u>appropriate</u> are the following responses by <u>Matthias</u>?

91. Write to the medical school and his personal tutor for advice as soon as possible.
92. Try to go to the hospital for 2 days to see if he can cope.
93. Do all of his book work whilst he is unable to walk around the hospital, so that he can focus on clinical training when he's better.
94. Stop going into hospital without letting anyone know.
95. Stop going into hospital and ask his friends to let the medical school know.

Scenario 20

Jessie, a medical student, has a friend called Gemma. Jessie suspects that Gemma has an eating disorder. Gemma was very stressed at medical school, and Jessie is uncertain with how to proceed.

How <u>appropriate</u> are the following responses by <u>Jessie</u> in this situation?

96. Ignore the situation and hope that someone else will notice.
97. Try to talk to Gemma and bring up her eating disorder.
98. Chat to Gemma about her stress and ask if she is coping. Allow her to bring up the disorder on her own account.
99. Speak to Gemma's parents about it, without consulting her.
100. Speak to your personal tutor for advice.

Scenario 21

Helen, a medical student, is has to retake her end of year exams because she failed them the first time. She had booked to go travelling with a friend in South America over the summer holidays, but the resit exams are during this period. She is worried that she will be unable to go and therefore, her friend will be left to travel by herself. If she goes she won't have enough time to revise for the exams.

How <u>appropriate</u> are the following responses by <u>Helen</u> in this situation?

101. Call her friend and cancel the holiday.
102. Go travelling but take her revision with her and revise every day.
103. Go for part of the holiday and come home early to revise for the exams.
104. Go travelling and cram in the revision in the few days between coming back and taking the exams.
105. Transfer her holiday booking to another mutually good friend so she can stay at home and revise but without leaving her friend to fend for herself.

Scenario 22

Daniel and Sean are medical students who are working together on a project. They get into a heated argument in the hospital lobby because Daniel has been prioritising his social life recently which is frustrating Sean.

How <u>important</u> are the following factors for <u>Sean</u> in deciding on what to do?

106. Sean can generally produce better work than Daniel anyway.
107. The mark that they get will be recorded in their log books.
108. Daniel and Sean have to work together for the rest of the year.
109. Daniel has recently broken up with his girlfriend.
110. Sean usually does most of the work when they have to do projects together.

Scenario 23

A medical student Tanya is invited to attend a clinic with Dr Garg who is in charge of Tanya's grade for the whole term. On the morning of the clinic, Tanya realises that she has not finished her essay that is due the next day.

How <u>important</u> are the following factors for <u>Tanya</u> to consider in deciding on what to do?

111. The importance of the essay towards her final mark for the year.
112. Tanya's friend did not find the clinic very educational.
113. Tanya's reputation with Dr Garg.
114. Whether or not Tanya will be able to attend a different clinic with Dr Garg.
115. How long it will take to finish the essay.

Scenario 24

Caroline, a final year medical student, is teaching first year medical students. She notices that they frequently arrive looking untidy and has noticed that some of the doctors have started to comment on how badly dressed the first year students are. She is worried that she will offend the students if she asks them to dress more appropriately, because 2 of the students are on her sports team.

How <u>important</u> are the following factors for <u>Caroline</u> in deciding on what to do?

116. Caroline's reputation with the doctors.
117. The first year students are only in hospital for 2 hours every week.
118. The first year students have direct contact with the patients and the hospital staff.
119. 2 of the students are on her sports team.
120. The first year students don't have their professionalism exams until third year.

Scenario 25

A medical student, Albert, is in his third year and is captain of the hockey team. He has noticed that his hockey training on Wednesday afternoons always clashes with his consultant teaching sessions. The consultant will be responsible for his final grade for the year.

How <u>important</u> are the following factors for <u>Albert</u> in deciding on what to do?
121. His hockey team needs him this year to win the championships.
122. Albert is on a sports scholarship at medical school.
123. His grade will determine if he can progress to his fourth year.
124. The consultant is free on Wednesday mornings.
125. This is the only teaching he will get on this particular topic this year.

Scenario 26

A patient is in a consultation with Dr Davison and Sybil, a medical student is observing. The doctor swears a number of times during the consultation, and Sybil notices that the patient is getting uncomfortable.

How <u>important</u> are the following factors for <u>Sybil</u> in deciding on what to do?

126. Dr Davison is marking one of Sybil's assessments.
127. The patient appears to be uncomfortable with Dr Davison swearing.
128. How often Dr Davison swears during other patient consultations.
129. If other members of staff are aware of Dr Davison's swearing.
130. If the patient has seen Dr Davison before.

Scenario 27

Marco and Alex are medical students on their surgical placement. They are invited to observe surgery with Mr. Daniels and told not to touch the sterile equipment. Before the operation begins, Marco sees Alex accidentally touch the sterile trolley with the operating equipment on it. Alex doesn't tell anyone, and Marco thinks that he should inform someone.

How <u>important</u> are the following factors for <u>Marco</u> in deciding on what to do?

131. The risk to the patient who is about to have the operation.
132. The inconvenience for all of the theatre staff to have to bring out a new sterile trolley.
133. Alex would be embarrassed because he touched something that he wasn't meant to.
134. Alex only briefly touched the trolley.
135. Mr Daniels would be disappointed in Marco and Alex.

Scenario 28

A medical student, Freddie, is on a busy hospital ward. A patient is addicted to pain medication and constantly bullies staff, so that they are reluctant to see her. Freddie has noticed that the doctors and nurses have been attending to her less frequently than before. One day, after she has been reviewed by the doctors and nurses, she starts to verbally abuse Freddie. She demands that he gets her more pain medication.

How <u>important</u> are the following factors for <u>Freddie</u> in determining what he should do?

136. The patient might be in pain.
137. The patient is being avoided by staff.
138. Freddie is not authorised to administer pain medication.
139. The patient may have already had her regular pain medication.
140. The patient has recently been reviewed by the doctors and nurses.

Scenario 29

Jenny is a junior doctor who is training under Mr. Gupta. Her sister, Claire, is due to be operated on by Mr. Gupta to correct a hernia. Claire is nervous about the operation and asks Jenny what she thinks about Mr. Gupta. Jenny knows that Mr. Gupta is a very good surgeon but he is often late when seeing patients on the wards. Therefore, he always appears to be rushing, flustered, and a little sweaty when speaking to them before their operations, which can make them lose confidence in him. Jenny must decide on what to say to her sister Claire.

How <u>important</u> are the following factors for <u>Jenny</u> in deciding on what to do?

141. Mr. Gupta is very competent.
142. The surgeon is often flustered when speaking to patients.
143. Mr. Gupta recently reprimanded Jenny for filling in a drug chart incorrectly.
144. Claire is already quite anxious about having an operation.
145. Patients usually do better if they are confident and at ease before an operation.

Scenario 30

A medical student, Alex, is on the university rugby team. He has been given model answers for various written assessments from older members of the rugby team. His friend, Annabel, realises that the Medical school often repeats their questions, and that it is against the rules to pass down previous papers from year to year.

How <u>important</u> are the following factors for <u>Annabel</u> in deciding on what to do?

146. Alex and Annabel dated for the first year, until he broke up with her.
147. Alex is a hard working student.
148. In this particular year, most of the questions won't be the same as in previous years.
149. Annabel isn't part of a team where information from older members is readily available.
150. The answers are not available to all students.

Scenario 31

Matthew, a medical student, is running late for his teaching session. He missed his bus and the next one doesn't come for another hour. His class was warned yesterday that if they were late for another session without a good reason, then they would not get a good grade for professionalism at the end of the term.

How <u>important</u> are the following factors for <u>Matthew</u> in deciding on what to do?

151. His marks for professionalism are not included in his final grade for the year.
152. Public transport information updates are readily available online.
153. The rest of his class is usually late.
154. Matthew is usually on time for most classes.
155. His teacher will be involved in his studies for the next year.

Scenario 32

Michaela, a medical student, is shadowing doctors on the intensive care unit of a busy hospital. Unfortunately Michaela becomes unwell during the week. She was told on her first day to stay at home if she becomes ill to minimise the risk of spreading infection to the patients. She is reluctant to remain at home because she this is her first and only week on the intensive care unit.

How <u>important</u> are the following factors for <u>Michaela</u> in deciding on what to do?

156. Michaela cannot spend another week on the intensive care unit.
157. Michaela's illness is just a mild cold and is unlikely to cause serious harm.
158. The illness doesn't affect her ability to interact with patients.
159. Michaela was told by the doctors not to come in if she became unwell.
160. Michaela's friend told her that she would learn a lot during her time on the intensive care unit.

Scenario 33

Jenny, a 4th year medical student, has booked and paid to go on the university ski trip. Unfortunately, she finds out that she has the option to sit a set of mock practical exams during the week of the ski trip. Jenny will lose the full amount of money if she pulls out of the ski trip. One of Jenny's friends has their mock exam the day after Jenny gets back. Jenny wants to try and swap their exam dates.

How <u>important</u> are the following factors for <u>Jenny</u> in deciding on what to do?

161. How useful the mock exams will be.
162. The cost of her ski trip.
163. The mock exams are optional.
164. The university's policy on swapping dates of exams.
165. Jenny has performed well on all exams in medical school so far.

Scenario 34

Luke, a medical student, has wanted to do a particular project for months. He has already spoken with the project's supervisor and planned it with him. He has also put it as his top choice for a project, although he knows that of other people also want to do the same project. Later next week, Luke finds out that he has been allocated to his second choice project instead. His friend, Architha, has been allocated the project that he wanted even though it was her last choice. Based, on this, Luke wants to make an official appeal.

How <u>important</u> are the following factors for <u>Luke</u> in deciding whether to appeal?

166. Luke has spoken to the project supervisor.
167. Architha didn't want to have the project.
168. The project grade will count for Luke's final grade at the end of the year.
169. Luke finds it difficult to invest time in a project that he doesn't care about.
170. Luke was allocated his second choice of project.

Scenario 35

Lucinda, a medical student, has been performing very well in her exams so far. She has been in a relationship with Andy (who is also in her year) for 6 weeks. Unfortunately, Andy will need to repeat the year as he has been struggling with the workload. Lucinda is desperate to stay in his year. Andy suggests that she fail her exams on purpose so that she can stay in his year.

How <u>important</u> are the following factors for <u>Lucinda</u> in deciding on what to do?

171. If Lucinda fails her exams the mark will be on her university transcript when she graduates.
172. Jobs as a junior doctor are partially determined based on your grades at medical school.
173. Lucinda and Andy have only been in a relationship for 6 weeks.
174. Lucinda's friends are performing very well and will progress onto the next year.
175. Andy has asked her to fail her exams on purpose.

Scenario 36

Shiv, a medical student, is nearly at the end of his rotation. He really wants to go to Australia for the Christmas holidays to see his girlfriend who is on her elective there. The flights are much cheaper if he skips the final day of his rotation. Most of his medical friends have already finished for the holidays because the doctors that were in charge of them finished their final assessments early. However, Shiv has a very strict doctor who insists that their final assessments will be on the final day of term, and no sooner. The doctor is due to retire after Christmas.

How <u>important</u> are the following factors for <u>Shiv</u> in deciding on what to do?

176. Shiv must pay for his own flights with the money he has saved up.
177. Shiv's final assessment involves the doctor asking him how the term has gone for him, and then signing his log book.
178. Shiv has been punctual and has produced impressive work throughout the rotation.
179. Shiv's doctor will be retiring after Christmas.
180. Shiv's girlfriend is in Australia.

Scenario 37

Jazzmynne, a medical student, is a talented vocalist and is offered the chance to go on a prestigious singing tour for a month with her choir. Unfortunately, this would mean missing a month of classes.

How <u>important</u> are the following factors for <u>Jazzmynne</u> in deciding on what to do?

181. Her choir has never been on an international tour before and this might be Jazzmynne's only chance to go.
182. Jazzmynne would end up missing half of one of her clinical rotations.
183. Jazzmynne's parents get anxious when she misses class.
184. Jazzmynne sometimes struggles to keep up with her workload.
185. Jazzmynne's friends are also going on the tour with her.

Scenario 38

Ellen, a medical student, has been writing for the university newspaper since her first year. She likes to focus on stories that are topical for the students. Recently, the new principal of the medical school has created a ban on stories that involve student bars and social lives. Ellen wants to start a petition to change this rule.

How <u>important</u> are the following factors for <u>Ellen</u> in deciding on what to do?

186. Ellen is in her final year of medical school.
187. The university has a good reputation for responding positively to student petitions.
188. 90% of the university newspaper readers are medical students.
189. The principal wants to encourage more stories about the health press, the world, and academics, rather than gossip at the bars.
190. Ellen doesn't like the new principal very much.

Scenario 39

Guy, a medical student, works at the student bar during the week. The recent change in the university's health and safety policy imposes a maximum number of students that can enter the bar at any time. This means that on popular nights, there is usually an hour-long queue to get into the bar. Guy thinks that this is unfair, and drafts a letter to the university, asking them to expand the student bar.

How <u>important</u> are the following factors for <u>Guy</u> in deciding on what to do?

191. Students are being denied access to their own bars.
192. Guy will get more work and therefore money if the bar expands.
193. The university has dropped in student satisfaction league tables.
194. The bar is very expensive.
195. The university is having financial troubles.

Scenario 40

Olivia, a medical student, wants to run for president of the student union. She has been involved for many years, and is very dedicated to the union. Her friend, Phil, also wants to run for president. Phil is very popular, although he has not contributed as much to the union as Olivia.

How <u>important</u> are the following factors for <u>Olivia</u> in deciding on what to do?

196. Olivia and Phil were previously in a relationship.
197. Phil is in his final year but Olivia is in her penultimate year.
198. Olivia has worked for the student union for a lot longer than Phil has.
199. Neither Phil nor Olivia are happy to run for any other position on the committee.
200. Olivia gave up the presidency of the student union last year because her friend who was in his final year wanted it.

Scenario 41

Annie, a final year medical student, is in the library. She notices George, another final year student, rushing out of the library and accidentally leaving some of his papers by the photocopying machine. She later finds out that they are copies of the upcoming final year exam.

How appropriate are each of the following responses by Annie in this situation?

201. Tell George she found the exam and won't inform anyone else as long as she can have a copy of the exam herself.
202. Copy the exam and distribute it to all the final year medical students so it is fair for everyone.
203. Ignore the situation and do nothing.
204. Alert her supervisor as to what she saw, and get advice on what to do.
205. Alert the medical school as to what she saw.

Scenario 42

Kate is a medical student on a ward round. A patient calls her and complains that a small amount of money has disappeared from her bedside table.

How appropriate are each of the following responses by Kate in this situation?

206. Call the police on the patients behalf, stealing is a criminal offense.
207. Tell the patient that she should have been more careful and you will see what you can do.
208. Tell the patient that she should not be making false accusations.
209. Ask the patient for further details about the theft, reassure her you will see what you can do and alert the nurse in charge of the ward.
210. Send an e-mail out to all other medical students to alert them of the theft.

Scenario 43

During a discussion with a fellow medical student, Andrew, in the student common room, Sam notices that a bag of marijuana falls out of Andrew's bag.

How appropriate are each of the following responses by Sam?

211. Accuse Andrew of drug possession and tell him he will be informing his supervisor.
212. Seek information from Andrew about why he has the marijuana in his bag, and tell him he will ignore the situation if he gets rid of it.
213. Discuss the incident with his own supervisor.
214. Offer support to Andrew and recommend he seek professional help.
215. Pretend he did not see anything and do nothing.

Scenario 44

Anna is a final year medical student attached to a medical team.. One morning, a junior doctor from the team arrives very drunk. The junior doctor is about to start his ward duties.

How appropriate are each of the following responses by Anna?
216. Alert a more senior member of the team to the situation at hand.
217. Do nothing as she is only a medical student.
218. Allow the junior doctor to carry out his ward duties and accompany him to make sure he is not making any mistakes.
219. Reflect on the situation later that evening.
220. Advise the junior doctor to inform the team and go home.

Scenario 45

Sean is a medical student working on a group project with three other students. One of the students, Sarah, is consistently arriving late and is not contributing her fair share of work to the group project.

How <u>appropriate</u> are each of the following responses by <u>Sean?</u>

221. Tell the group's supervisor about Sarah's tardiness and contribution.
222. Discuss the problem with other members of the group.
223. Approach Sarah and tell her that her attitude and poor contribution are causing problems and she needs to pull her weight. .
224. Ask Sarah if there is a reason for her lateness and lack of contribution and if there is anything Sean can do to help.
225. Ignore the situation. The rest of the group is working well and you can work harder to complete the group project.

Scenario 46

Mary is a medical student who is attending a placement on the wards. One day she sees a nurse pick antibiotics out of a drug trolley and place them in her handbag for personal use.

How <u>appropriate</u> are each of the following responses by <u>Mary?</u>

226. Approach the nurse and inform her that this is not best practice.
227. Inform the nurse in charge.
228. Report the matter to the Consultant supervising her.
229. Pretend she did not see anything as the nurse probably needs the antibiotics.
230. Write a reflective piece on the nurse's practice.

Scenario 47

Alan is a medical student who is having a bedside teaching session.During the session the Consultant asks Alan a range of questions, some of which he struggles to answer. At the end of the session Alan feels that the Consultant was rude and has embarrassed him in front of the patient.

How <u>appropriate</u> are each of the following responses by <u>Alan</u> in this situation?

231. Complain about the incident to a senior nurse.
232. Arrange a meeting with the Consultant to discuss the incident.
233. Argue with the Consultant at the bedside so he knows he is being rude.
234. Arrange a meeting with his medical school supervisor to discuss the incident.
235. Once the session is over apologise to the patient about the consultant's behaviour.

Scenario 48

Harry is a medical student doing a placement on a medical ward. One of the nurses comes up to Harry and complains to him about one of his colleagues, Joe's, body odour and is asking if he could have a quiet word with him.

How <u>appropriate</u> are each of the following responses by <u>Harry</u> in this situation?

236. Tell the nurse that this is really an issue for the ward Consultant to deal with and she should go and talk to him,
237. Raise the issue in confidence with Joe.
238. Tell the nurse he will speak to Joe about it, but then ignore the issue. Harry does not want to hurt Joe's feelings.
239. Send an anonymous note to Joe.
240. Raise the issue at a student group meeting with Joe present.

Scenario 49

Helen is a final year medical student. She has written a case report for publication and her Consultant has recently reviewed the final draft. When he gives her the case report back with comments, he has added two names to the list of authors. Helen enquires and the Consultant explains that it is his wife and his Registrar who are currently applying for jobs and need publications on their CVs.

How <u>appropriate</u> are each of the following responses by <u>Helen</u> in this situation?

241. Tell the consultant that she can't publish the case report with the two names added and will only publish it under her name.
242. Agree to add the names as it's only a case report and not a research paper.
243. Discuss the matter with her medical school supervisor.
244. Discuss the matter with the consultant in a private meeting.
245. Choose not to publish the case report at all.

Scenario 50

The medical school library has sent out a notice for a few missing books. John and Clare, both medical students, are studying at her home. Whilst they are studying, John sees a pile of books on her shelf which suspiciously look like the ones from the library.

How <u>appropriate</u> are each of the following responses by <u>John</u> in this situation?

246. Accuse Clare of stealing the books from the library.
247. Ask Clare if these are actually the books from the library and tell her to return them if they are the missing books.
248. Contacting the police, theft is a crime.
249. Inform the medical school library of his suspicions.
250. Discussing the situation with his supervisor.

Scenario 51

Nadia, a fifth year medical student, is currently on a four week rotation on the neonatal wards. She has been assigned to shadow one of the consultants, Anne. During her first week, whenever Nadia approaches Anne, she is told that 'there is not much going on today, why don't you head home early'. This continues for a week and Nadia is starting to feel like she hasn't learnt anything whilst shadowing Anne.

How <u>appropriate</u> are each of the following responses by <u>Nadia</u> in this situation?

251. Speak to Anne about her feelings of being ignored, and find a solution with Anne.
252. Watch and wait- there is still three weeks left to her attachment and things may improve.
253. Discuss the situation with her supervisor.
254. Stop coming to the ward and go to the library instead.
255. Find another doctor on the ward who she can shadow.

Scenario 52

Dean,a third year medical student, is currently on a placement in hospital with a group of three other students. During their placement they have to get their attendance and some clinical skills signed off by doctors. On the last day of their placement, one of the other students, Laura, approaches Dean and asks him to sign off some of her attendance and two of her skills.

How <u>appropriate</u> are each of the following responses by <u>Dean?</u>

256. Reprimand Laura for delaying signing the book until now, refuse to sign the book and give her tips of how to be more organised.
257. Agree to sign the missing boxes for Laura, Dean knows she attended everyday.
258. Suggest she finds one of the junior doctors to sign her book.
259. Ask Laura to demonstrate the skills and then sign them.
260. Report Laura to the medical school.

Scenario 53

Emma is shadowing a junior doctor, Steven, on the wards. He is trying to cannulate a patient and after three failed attempts, he stops and leaves the patient's bedside flustered without an explanation. Emma is still stood by the patient's bedside.

How <u>appropriate</u> are each of the following responses by <u>Emma</u> in this situation?

261. Try to cannulate the patient herself, she was successful in inserting one yesterday.
262. Speak to the nurse in charge about Steven.
263. Excuse herself from the patient's bedside and go and find Steven.
264. Speak to Steven's consultant.
265. Complain about Steven to the patient.

Scenario 54

Jill is a medical student. She is shadowing a registrar in surgery, who is currently clerking a patient who needs an operation. The registrar leaves Jill with the patient and goes to get the operation consent forms. While he is away the patient makes a racist comment about the registrar.

How <u>appropriate</u> are each of the following responses by <u>Jill</u> in this situation?

266. She should retaliate and call the patient a racist idiot.
267. Pretend she didn't hear the patient's comment.
268. Politely tell the patient that this language is not tolerated in the hospital.
269. Discuss the matter with the consultant in charge.
270. Report the matter to the registrar.

Scenario 15

John is a first year medical student, he is working with four other medical students on a group project. Hannah, one of the other students, is dominating the group discussions and is doing a large amount of the work.

How <u>appropriate</u> are each of the following responses by <u>John</u> in this situation?

271. Discuss the issue with the other members of the group.
272. Speak to Hannah and tell her she is not allowing others to contribute ,and if she doesn't change she has to leave the group.
273. Allow Hannah to continue to dominate the group discussions and to do the majority of the work.
274. Speak to his supervisor.
275. Speak to Hannah and tell her it would be good if all members of the group contributed equally to the project.

Scenario 16

Jenny, a medical student, hears that one of her colleagues has recently lost their grandmother. They are both at hospital on their rotation, when her colleague bursts into tears.

How <u>appropriate</u> are each of the following responses by <u>Jenny</u> in this situation?

276. Advise her colleague to go home for the day.
277. Offer her colleague support.
278. Discuss this with her supervisor.
279. Tell her colleague it is inappropriate to cry on the wards and to stop.
280. Give her colleague some time alone.

Scenario 57

Maya, a sixth year medical student, is out with a group of her friends from medical school. They are all out at a restaurant for an evening meal. The students start discussing some of the interesting patients they had seen over the last few weeks. Some of the patients' names are being mentioned in the conversation, and other diners could certainly overhear.

How <u>appropriate</u> are each of the following responses by <u>Maya</u> in this situation?

281. Maya should join in the conversation and speak about a particularly interesting patient she had seen.
282. Maya should warn her friends that they are in a public place and should not be talking about patients using their names.
283. Maya should try to change the subject of the conversation.
284. Maya should report her friends to the medical school
285. Excuse herself from the table and leave.

Scenario 58

Michael is a medical student; one of his colleagues Simon confides in him that he is suffering from depression and feels it is affecting their work. He asks Michael not to tell anyone.

How <u>appropriate</u> are each of the following responses by <u>Michael</u> in this situation?

286. Speak to his own supervisor without telling Simon.
287. Advise Simon to speak to his supervisor.
288. Mock him and tell him he should get over it.
289. Tell other students in the medical school so they can offer some support.
290. Offer any help he can, and advise him to see his GP or a counselor.

Scenario 59

Mark is a medical student, he walk into the doctor's mess and sees one of the doctors watching adult pornography on one of the hospital computers.

How <u>appropriate</u> are each of the following responses by <u>Mark</u> in this situation?

291. Pretend he didn't see anything and leave the doctor's mess immediately.
292. Speak to his supervisor.
293. Confront the doctor about the issue.
294. Spread the news in the hospital that there was a doctor watching pornography during working hours.
295. Contact the hospital IT services.

Scenario 60

Alice has arranged to go out for dinner with her boyfriend tonight. Just before leaving medical school she receives an e-mail which informs her that tomorrow morning's seminar has been moved to this evening and starts in the next 30 minutes.

How <u>appropriate</u> are each of the following responses by <u>Alice</u> in this situation?

296. Tell another student to apologise to the seminar leader for her absence.
297. Leave as planned, and hope her absence goes unnoticed.
298. Find out what the seminar is about and speak to the seminar leader and explain that she had prior arrangements.
299. Call her boyfriend and tell him that the dinner needs to be cancelled, as she must attend an important teaching session.
300. Write a rude email to the seminar leader complaining about the short notice.

ANSWERS

Verbal Reasoning Answers

Q	A	Q	A	Q	A	Q	A	Q	A
1	False	51	False	101	False	151	B	201	D
2	False	52	False	102	True	152	A	202	D
3	True	53	True	103	False	153	A	203	C
4	Can't tell	54	True	104	False	154	C	204	B
5	Can't tell	55	Can't tell	105	False	155	C	205	A
6	False	56	False	106	Can't tell	156	B	206	A
7	False	57	True	107	B	157	A	207	C
8	Can't tell	58	False	108	C	158	D	208	A
9	Can't tell	59	False	109	B	159	B	209	C
10	True	60	True	110	B	160	A	210	A
11	False	61	False	111	D	161	C	211	A
12	False	62	Can't tell	112	C	162	C	212	C
13	Can't tell	63	False	113	B	163	B	213	A
14	False	64	True	114	D	164	B	214	C
15	Can't tell	65	False	115	C	165	A	215	D
16	False	66	False	116	D	166	D	216	B
17	Can't tell	67	False	117	A	167	A	217	B
18	True	68	False	118	C	168	C	218	A
19	Can't tell	69	Can't tell	119	C	169	D	219	A
20	True	70	True	120	A	170	C	220	C
21	Can't tell	71	False	121	C	171	C	221	True
22	Can't tell	72	False	122	C	172	A	222	D
23	True	73	False	123	D	173	B	223	C
24	False	74	False	124	B	174	A	224	C
25	True	75	False	125	A	175	A	225	B
26	False	76	False	126	D	176	C	226	C
27	True	77	False	127	B	177	B	227	D
28	False	78	True	128	C	178	C	228	A
29	Can't tell	79	False	129	C	179	B	229	C
30	False	80	False	130	D	180	D	230	D
31	Can't tell	81	Can't tell	131	C	181	C	231	C
32	False	82	Can't tell	132	D	182	C	232	A
33	False	83	Can't tell	133	B	183	B	233	B
34	False	84	Can't tell	134	C	184	A	234	D
35	False	85	False	135	C	185	B	235	C
36	False	86	True	136	C	186	C	236	False
37	Can't tell	87	Can't tell	137	B	187	A	237	C
38	Can't tell	88	True	138	B	188	C	238	A
39	False	89	False	139	D	189	D	239	C
40	True	90	Can't tell	140	A	190	A	240	A
41	True	91	False	141	B	191	A	241	True
42	Can't tell	92	False	142	C	192	C	242	False
43	False	93	False	143	C	193	D	243	Can't tell
44	False	94	False	144	C	194	A	244	Can't tell
45	True	95	True	145	C	195	C	245	False
46	Can't tell	96	Can't tell	146	C	196	A	246	False
47	False	97	True	147	A	197	B	247	Can't tell
48	True	98	True	148	C	198	C	248	Can't tell
49	True	99	False	149	C	199	C	249	True
50	Can't tell	100	True	150	B	200	B	250	True

Set 1:

1. **False** -In paragraph 2, the passage says that only the countries that signed the protocol were legally bound.

2. False -In 2004, the condition that over 55% of emissions were accounted for was met without Australia and the US.

3. **True** -in paragraph 2: 'each country that signed the protocol agreed to reduce their emissions to their own specific target'.

4. **Can't tell** -We know this is when the Kyoto Protocol was enforced but there is no information to suggest whether emissions actually decreased.

5. **Can't tell** -Paragraph 4 says that a 60% emission reduction would have a 'significant impact', but we cannot tell if the remaining effects would be harmful or not. The passage is also talking about a global reduction of 60% - there is no mention of each individual country needing reduce their emissions by 60%.

Set 2:

6. **False** -From paragraph 1, we can see that the Soviets were mainly concerned with showing off their economic power and technological superiority.

7. **False** -Paragraph 3 says that Project Apollo was tasked with landing the first man on the moon.

8. **Can't tell** -The passage does not tell us why the Soviets did not land a man on the moon.

9. **Can't tell** -We do not know the state of the American space efforts prior to the launch of Sputnik – we just know that its launch 'prompted urgency'.

10. **True** -The Soviets had just sent Yuri Gagarin into space and it wasn't expected that the US would beat the Soviets in landing a man on the moon; therefore the US was behind the Soviets at this stage.

Set 3:

11. **False**- Paragraph 1 says Pheidippides was the fastest runner

12. **False**-Paragraph 2 says that they were standardised from 1921 but using the 1908 distance

13. **Can't tell**-Paragraph 1 says the Greeks were not expecting to beat the Persians, but we do not know if the Persians were expecting to beat the Greeks, despite their larger army

14. **False**- Paragraph 2: to be an IAAF marathon the distance must be 26.2 miles. The original route was 25 miles

15. **Can't tell** -We know that the Persian soldiers outnumbered the Greeks and had superior cavalry, but we are not told about their training

Set 4:

16. **False**- From paragraph 1, we can see that birds would not survive if they did not migrate

17. **Can't tell**-Whilst there is only mention here of migrating in flocks, nowhere does it specifically state that either all migrating birds do so or that any migrating birds do not

18. **True**- Although the leading bird does not benefit at the time, they continually change position within the flock according to paragraph 3

19. **Can't tell** -Paragraph 3 shows that the Northern bald ibis does not behave in this way so it is not known what would happen

20. **True** -Paragraph 1 states they migrate north in the spring, and south in the winter

Set 5:

21. **Can't tell** -We know the amount of wild flowers has decreased by this amount but we can't tell if the bee population fell by a proportional amount. The 50% reduction mentioned in Paragraph 1 only refers to honey bees

22. **Can't tell**-Whilst the UK have experienced a significantly greater decline than the European average, there is no data from the rest of the world.

23. True -Paragraph 2 tells us modern techniques produce more food.

24. **False** -As well as pollination, bees are also involved in complex food chains and cannot be replaced

25. **True** -From paragraph 1 we see that pesticides are one of the causes of bee population decline. Therefore, reducing pesticide usage would reduce this rate of decline.

Set 6:

26. **False** -This would affect the thinking distance, not the braking distance.
27. **True** -From paragraph 2 we see that the increased traction of winter tyres is because the softness of their material allows better traction.
28. **False**- From paragraph 4, we are told it is not safe to run the engine when the exhaust pipe is blocked because of the risk of carbon monoxide production.
29. **Can't tell**-There is not sufficient information in the passage to suggest whether the softer tyres are dangerous to use in hot conditions.
30. **False**- Paragraph 3 says it is safer to steer out of the way than to brake.

Set 7:

31. **Can't tell** -In Paragraph 1, we are told the method is named after Socrates, but we are not told whether he was the first person to use it or whether he simply adapted and/or popularised the Socratic method.
32. **False** -The Socratic method challenges the statements and opinions that a person makes, not their wisdom.
33. False -Although Socrates was trialled with these charges, the Socratic method as described was not the cause.
34. **False** -We can see from the final paragraph that he chose to die defending knowledge and wisdom, rather than fleeing in fear.
35. **False** -From paragraph 2 we see that Socrates did not provide answers to these difficult questions.

Set 8:

36. **False** -Paragraph 1 says the dialects of the Saxon, Angle and Jute tribes formed the Anglo-Saxon language.
37. **Can't tell**-Paragraph 2 says English speakers would struggle to understand Old English, but we do not know about German speakers.
38. **Can't tell**-The story of Beowulf only mentions that he kills Grendel and his mother, not about his strength compared to other warriors.
39. **False**-The Friesian dialect is similar to Old English but they are not the same.
40. **True** -The Celtic Britons were originally living in England before being forced into Wales.

Set 9:

41. **True**-Paragraph 1 says that stars in the constellation Cassiopeia are visible only in the northern hemisphere; therefore they are not visible in the southern hemisphere.
42. **Can't tell** -The passage does not tell us if Cassiopeia rises and falls in the sky or not.
43. **False**-Paragraph 3 says the star sign represents the position of the sun, not the visible constellations.
44. **False** -Paragraph 2 says Polaris is used for navigation as its position is constant in the sky.
45. **True**-Paragraph 1 mentions different constellations being visible from different places on the Earth.

Set 10:

46. **Can't tell**-The passage tells us about maple tree sap but we don't know about other trees.
47. **False** -He was angered because of the laziness of the village people.
48. **True** -Paragraph 1 says that the process is similar although different equipment is used.
49. **True**-Paragraph 1 says the sap is only harvested during March.
50. **Can't tell** -Although the laziness associated with drinking maple syrup suggests hunting is harder than syrup drinking, the people may just have not hunted because they did not need to – not because it was harder.

Set 11:

51. **False**-Paragraph 1 says Stockholm syndrome happens sometimes, not always.
52. **False** -The passage suggests that the hugging and kissing occurred due to the positive relationship formed between the captors and hostages.
53. **True** -Paragraph 4 says the FBI is willing to devote resources to understanding Stockholm syndrome in order to aid crisis negotiation.
54. **True**-The explanation given in paragraph 3 suggests that the hostages fearing for their life is a step in the development of Stockholm syndrome.

55. **Can't tell** -Although this incident is the origin of the name 'Stockholm syndrome', it is not mentioned whether this phenomenon has happened before

Set 12:

56. **False**-It is this mucous layer that offers protection from the anemone tentacles

57. **True**- Paragraph 2 mentions that the bright colour of the clownfish lures in fish which the anemone eventually stings and eats.

58. **False**-Paragraph 3 mentions that anemones increase the lifespan of clownfish but are not essential for clownfish to live.

59. **False**- Paragraphs 1 and 3 mention that only certain anemone form this mutually beneficial relationship.

60. **True** -Paragraph 2 says the anemone sting protects the clownfish from predators.

Set 13:

61. **False**-Paragraph 2 indicates he was opposed to the death penalty.

62. **Can't tell**-The passage only tells us that the guillotine was commonly used during the French Revolution, it doesn't tell us what was used for execution before the French Revolution.

63. **False** -Although it can cause death by asphyxiation, the common cause of death is snapping of the neck.

64. **True**- This was Joseph Guillotin's explanation.

65. **False** -Paragraph 1 tells us the guillotine was the common form of execution as it was the only legal method of execution.

Set 14:

66. **False** -The Gherkin is famous for its distinctive shape, and although this forms part of a cooling system it is not famous for that reason.

67. **False** -The City of London governing body wanted redevelopment to restore the old historic look – implying the Baltic Exchange had this look before the bombing.

68. **False** -The Gherkin was bought for £630m and sold for £700m, so a profit was made on the sale.

69. **Can't tell** -We are told the building is damaged by bombs, but we are not told what the intention of the attack was.

70. **True**- It is stated in paragraph 1 that Norman Foster is famous for this.

Set 15:

71. **False** -The passage tells us London is significantly more expensive.

72. **False** -There are no legal implications for paying below the living wage as it is the minimum wage that is legally binding.

73. **False**-This individual would be earning less than the living wage and so would not be able to live comfortably according to the definition in paragraph 1.

74. **False** -Some employees will be fired and others would have to work harder; some employees may already be earning more than the living wage and so would not benefit.

75. **False** -The family of that individual will also benefit from the living wage, as well as the company they work for.

Set 16:

76. **False** -Although a country with more people will likely have a greater ecological footprint, it is through their increased usage of resources, not a direct consequence of the population. There can be significant discrepancies.

77. **False**- This is the bio-productive capacity of the Earth, not the total space available.

78. **True** -The final paragraph mentions that reducing use of unsustainable resources will reduce the ecological footprint.

79. **False** -The final paragraph says that a global effort is needed.

80. **False** -The final paragraph says that we can increase the bio-productive space available.

Set 17:

81. **Can't tell** -Paragraph one states that some believe them to be 'one of the most important', which does not necessarily mean they are the most influential.
82. **Can't tell** -The passage states it started the UK punk movement, but this does not necessarily mean it began the global movement.
83. **Can't tell** -The passage states that there were lyrics written about abortion, but they do not state the moral stance these lyrics took.
84. **Can't tell** -The passage tells us that some songs attacked the music industry and that controversial topics such as the Holocaust were commented on, however we do not know if these two topics were said to be related.
85. **False** -They attacked 'blindly accepting royalty as an authority', and so attacked a royalist standpoint.
86. **True** -Not at the same time, but The Sex Pistols had two individual bassists over the course of their existence.

Set 18:

87. **Can't tell** -The passage doesn't mention what proportion of people exposed to asbestos will get mesothelioma.
88. **True** -As breast cancer is the most common cancer in the UK and very rare in men, it must be very common in women.
89. **False** -The success rates of tumour removal procedures were improved by surgical hygiene.
90. **Can't tell** -The passage doesn't mention the relative success rates between the two methods.
91. **False** -Some tumours are able to metastasise, meaning that there are also tumours that do not.

Set 19:

92. **False** -The land was claimed because the British people did not believe that anybody actually owned the land.
93. **False** -Not all of the 517,000 Aborigines are living in cities and towns.
94. **False**-They were only able to reclaim the land that they could prove they originally owned.
95. **True** -Only a few Aborigines lack the education to integrate with modern Australian life.
96. **Can't tell** -We do not know what the south-east Asian lifestyle involved.

Set 20:

97. **True** -It was discovered that chilli peppers had similar taste to black peppercorn and so became valuable around the world.
98. **True** -Paragraph 2 says that spicy food allows people to sweat.
99. **False** -Paragraph 1 says 'most' peppers are spicy, but not all.
100. **True**-The capsaicin receptor is responsible for the pain associated with spicy chillies.
101. **False** -The description of labelled line coding says these different methods of activating this receptor will cause the same sensation.

Set 21:

102. **True**- Dropping out of school will affect the level of education, which is a social determinant of health.
103. **False** -They can also be measured by the quality of life.
104. **False** -Although this would prevent the described mechanism from happening, many other mechanisms are acting at the same time. This was just one example of many possible mechanisms.
105. **False** -The final paragraph says we are reducing the gap between those more privileged and those who are more disadvantaged.
106. **Can't tell** -The life expectancy described in paragraph 2 is an average and so can't reliably be used in specific examples.

Set 22:

107. **B-** The passage suggests that the critics suppose 'human beings to be capable of no pleasures except those of which swine are capable', and so suggest the potential of higher pleasures. They do not call their critics degraded, but suggest that the critics degrade human nature, nor do they accuse critics of being either miserable or indulgent.

108. **C-** This action does not see happiness as its justification, but conforming to social norms. It does not seek to increase pleasure or reduce pain, just follow dogma, which is not according to the principle of utility.

109. **B-** Mill does not describe the religious views of his critics, or explicitly call them reactionary, and in fact describes them 'some of the most estimable in feeling and purpose'. He does mention the countries of origin for some of his critics, and they are all in Europe.

110. **B-** The passage also states 'desirable things' could be 'means' to pleasure.

111. **D-** Utility is defined as the foundation of the system, and a belief that the increasing of happiness/decreasing of happiness is good. The passage specifically states that it has not given an exhaustive definition of things that are pleasurable/painful, and it does not specifically define Epicureans, simply suggests a link between them and utilitarian's - which is not an exhaustive explanation of the term.

Set 23:

112. **C-** Though sandstone is made from sand, the passage does not state that ALL rocks are made from this material.

113. **B-** The passage discusses the valley when implementing the wall-building analogy.

114. **D-** 'Some ancient source' is all we are told, and so the source is undisclosed (Paragraph 3).

115. **C-** This rock is said in paragraph 2 to be found 'everywhere', so not 'nowhere'.

116. **D-** Paragraph 3: The grains are said to be sorted in groups 'of a size', i.e. measurements. They are all described as 'worn and rounded', and no mention is made of differentiation through age/shape.

Set 24:

117. **A-** Though the flowers smell pleasant, no mention is made of this scent being used to manufacture perfumes.

118. **C-** The passage states that the flower could bloom more than once a century, precluding 'D', but that it is thought to only do so in a century, providing evidence for 'C'.

119. **C-** The statement claims that Narcissus plants are 'prized by many' over Lilies, but this does not mean that all people - or even the majority of people - think the former is more attractive/better than the latter, or that all homeowners enjoy the plant.

120. **A-** It is actually a substance 'very similar to rum', not rum itself.

121. **C-** The genus 'belongs' to the family, as stated in paragraph 1.

Set 25:

122. **C-** Women in Illinois, not across USA, were subject to the law, and the passage does not state either a change in fashion or actual arrests, only the potential for arrests.

123. **D-** The pulling out of feathers from live birds was seen as the negative to using osprey feathers.

124. **B-** They could be possessed only 'in their proper season'.

125. **A-** The problem cited is that the article was already in use in the clothing of numerous military men. The authority of the princess/sexist politics does not feature in the passage, and 'D' is patently false

126. **D-** None of those are precluded, as only 'harmless' and 'dead' birds (in their entirety) were prohibited. Wearing a living bird was not explicitly banned.

Set 26:

127.**B**- Nothing in the above passage provides evidence for 'C', 'D' or 'A' - in fact, the 'indie' description of Pope opposes the idea of him working for a games company at all. 'B' is supported by the fact the game is available on a number of devices and system operators.

128.**C**- The immigration officer's job is to process people correctly - not to grow or limit the number of immigrants, as in 'A' and 'B'. 'D' is vague, as 'to stamp passports' does not necessarily mean to stamp them correctly, and false stamps would be counter to the purpose of the border guard's position.

129.**C**- Though the game player may perform either a body scan or finger print check when something is amiss is the candidate's documents, they will not necessarily do either of these - first, they will 'enquire', which is synonymous with 'C', 'asking...for further information'. The only one of the statements that will be universally true for discrepancies therefore is c.

130.**D**- The game-player may accept 'bribes', so 'A' is not true. The game player 'may' arrest candidates, but the passage does not state he or she must, so 'B' is false. The game-player is allowed two mistakes, so to an extent, can be forgiven - making 'C' incorrect. As further mistakes will lead to being 'pecuniarily punished', 'D' is the only accurate statement.

Set 27:

131.**C**- The head of LA NAACP is a civil rights' activist.

132.**D**- The other aspects may appear in films, but only racial slurs were cited as a 'common' element without specifying location of setting of sub-type.

133.**B**- 'Primarily' means the same as 'predominately' in this case, and the cast has been described as 'predominately' black - meaning most, but not all, cast members are black.

134.**C**- Original intended audiences were black city-dwellers: 'B' is too broad and 'A' is too narrow. 'D' is not at all supported in the passage. 'C' describes the growing audience, and how the genre is now 'not exclusive to any race' (Paragraph 2)

135.**C**- It is possible, but nothing in the statement suggests the potentially racy titles are a nod towards sexploitation genre. The innuendo may be incidental, or the choice of words completely divorced from the pornographic films of the past.

Set 28:

136.**C**- It would be a massive assumption to state that just because two characters in a book are 'vicious', all of them will be, so 'A' is not necessarily correct. 'B' also believes in a despair that is described to belong to the Comedian, but not Rorscach. The argument of the passage is that 'D', which Moore may believe, is not the case - the beloved character is not simply worshipped for his violence, but for his belief in justice. 'C' is correct, as Moore describes how he wished Rorschach not to be a favourite character, but a warning.

137.**B**- False. The fact he finds things a 'joke' is what makes him the Comedian. He may not be all that funny, but the joke is still there.

138.**B**- He does not mention madness ('D'), or invoke shame ('C'), or simply state it is good to be good ('A') - specifically, he states we must act as if the world is 'just', even when not, to attain dignity.

139.**D**- No value judgment is made comparing violent actions or on the Comedian's jokes so 'A' and 'B' are false. The passage also acknowledge Rorschach's violence, showing 'C' is wrong, but does state that his actions are due to the fact he believes he is acting in the name of justice, which lends him an ethical justification to his actions.

140.**A**- The lack of meaning in anything is what leads him to treat everything as a joke - it is not hatred, but the inability to see 'purpose' in himself or his fellow man.

Set 29:

141. **B**- The passage's explanation of The Bechdel Test does not state that the test is used to show 'sexism', so a failed film does not equate to a sexist one.

142. **C**- This is the only film that shows a female-female conversation not on a man. 'A' and 'B' only feature conversations focused on males and 'D' does not have two women speaking to one another.

143. **C**- Though of the two horror films mentioned in the extract one passes it and another one fails, the passage does not make use this to make any claims on the genre as a whole.

144. **C**- Her comment that it is 'strict' shows she has a reservation, but her general agreeing that it is a 'good idea' shows she approves of the notion. Her statement is too qualified to be described by 'A', too positive to be understood by 'B', and contains a judgement precluding 'C' as a correct statement.

145. **C**- The two women are not named within the comic strip - though two women are behind the idea (one in suggesting it, the other in illustrating it), that is not to say that the two depicted women are the same.

Set 30:

146. **C**- The possible 'idea' that repetition is funny is inherently funny is mentioned, but neither confirmed nor denied.

147. **A**- This is the only statement actually mentioned above. Laughing twice at the same joke does not make it 'twice as funny', that is a logical fallacy; it does not state that all relationships dictate that the two parties find themselves funny (this is too vague) and being 'comfortable' is not cited as encouraging laughter.

148. **C**- There is no mention of a humour requirement for the first joke.

149. **C**- Though the call-back here is described as a comic trope, this does not necessarily preclude the same device being used in something un-comic: in the same way word-play or alliteration can be used to achieve a multitude effects, not all humorous.

150. **B**- Though 'A' is a potential, it is not a requirement of the call back, and 'C' is not supported through the passage's material. At no point is the call-back named significant, compared to other tropes: it is simply one that the passage focuses on. It does however state that it can be used by a comedian, and used for comedy, so 'B' is correct.

Set 31:

151. **B**- She was the fourth, after Catherine, Mary and her namesake Harriet Elizabeth, who died as an infant.

152. **A**- He was a 'Rev', a reverend, and a 'divine'. She was born in the USA, not UK, and in the 19th century. Her town is said to be 'characteristic', and so not 'average'.

153. **A**- She is said to have 'veneration' in all who knew her.

154. **C**- It is only described as the 'most sad' and 'most tender' memory of her childhood, not her life. It is, however, described as 'the first memorable incident' and thus the earliest one of her life.

155. **C**- She wrote her Charles, so at least wrote one letter. The autobiography mentioned belonged to her brother, not her, and she had five brothers and sisters waiting for her when she was born, not only brothers. She was actually four when her mother died, as the narrator tells us, and her remembering being between 'three and four' is a false memory.

Set 32:

156. **B-** He thought it was 'a pity that only rich people could own books', and from this he 'finally determined to contrive' of a new way of printing. The passage does not state that he wished to make money, found books too expensive to get a hold of or was impatient himself when it came to the production of books.

157. **A-** The need to be careful is mentioned, as is the fact that the process takes a long time both in creating the block and due to the fact one block can only print one page. That it may tire a carver to make the block is possible, but it is not cited in the passage.

158. **D-** The statement says it is 'very likely' he was taught to read, but is not definite. That his father comes from a 'good family' does not mean he is a member of the aristocracy, necessarily. Though block printing was used as the boy grew up, it does not state this was the most popular process. The mention of Gutenberg's family's 'wealthy friends' indicates they were sociable.

159. **B-** The paper was laid on top of the block, not underneath.

160. **A-** There is nothing written in the passage praising the craftsmanship of manuscripts. The appropriateness of the titles for both book production processes is explained, and the 'wealthy friends' are described as a source to borrow books from, thus a way of expanding your reading.

Set 33:

161. **C-** In both examples given above, Apollo is the giver - but these are only two of 'several' versions, and it is possible others exist.

162. **C-** The gift of prophecy is supernatural, therefore this is the best answer. There is no evidence to suggest that she knew chastity would lead to her tragic fate, that she is 'often' seen as a home wrecker (though she is killed by Clytemnestra, this does not prove Cassandra, the slave, is perceived as a vindictive 'other woman') or that parents use the name to express hate – they may just like the sound, and not care about its history, or not know the history at all.

163. **B-** Nothing is said about the state of Cassandra's childhood, but it is mentioned that plays are written about her and that Homer has written about her, showing 'C' and 'D' are true. The fact her father was a king proves she was from a royal line.

164. **B-** The passage states the 'presentation of her character alters'.

165. **A-** It is not said that she ignores or refuses to accept the gift, or demands more presents, even in the first story, just that she refuses to sleep with him. Only the broken promise, as described in the second version of the myth, is mentioned above.

Set 34:

166. **D-** The passage attacks a generalisation, and shows an example that refutes one given to the 'musical' genre. Nothing is mentioned of Sondheim's talents, or what his role was in creating the musical, nor are their claims made to Wheeler's literary tastes (he may just like ONE penny dreadful). This musical may deal with morbid themes, but that's not to say that most do - it could be only a select few that do.

167. **A-** The pies make the crimes 'culinary' in nature, the mention of revenge shows Todd's illegal acts to be 'vengeful' and the judge's rape is a 'sexual' crime. There is nothing explicitly suggesting the crimes of any party are funny, or to be considered funny.

168. **C-** Though the original title 'A String of Pearls: A Romance' may appear to suggest a romantic relationship within the narrative, nothing in the passage states the two are a couple.

169. **D-** Is essentially synonymous is the quoted belief, 'we all deserve to die', which include both bad and good people and makes no significant reference to gender exclusion/inclusion.

170. C- There are four mentioned themes, but that does not mean there are only four themes, nor does 'legal corruption' get named as the central theme. As the entirety of Sweeney Todd is not discussed in the passage, only a central plot line, one could not exclude the potential of something positive happening in the play - even a minor incident. The themes mentioned are, however, indeed macabre.

Set 35:

171. **C-** The above passage is about WWII trains, not WWI ones.
172. **A-** The soldier makes a chair by using a tipped-up suitcase. It is a marine, not a sailor, learning on the back of a chair, and the passage states 'some', not 'many' queue for two hours to go to the diner car, whilst similarly 'some' (not necessarily 'many') go hungry.
173. **B-** The passage discourages mothers from going on trains with a baby, stating they should only do it as a necessity.
174. **A-** 'At every stop' more people come on the train.
175. **A-** The passage does not insult anyone, but it does say the railroads are doing their job 'well'.

Set 36:

176. **C** The passage states the slice of bread is an ounce, and contains 3/4 ounce of flour, making it 75% flour.
177. **B-** It says it is possible that they do, but also possible that they don't. There is no emphatic claim.
178. **C-** It is over a million loaves a day, 319 million pounds - not bushels - a year and 365,000 loaves - not over - a year.
179. **B-** False. It may seem unimportant, but the passage goes on to explain how a single slice of bread has value.
180. **D-** The government have researched waste, but not taken responsibility for it, nor is it said they should do more to combat this. The final sentence confirms that housekeepers are responsible for their own food wastage.

Set 37:

181. **C-** Fourteen women and three men are described to be arrested.
182. **C-** She was the only woman, but it does not state whether either man was brought to trial.
183. **B-** The majority ultimately rule, but Hall 'dissented', or disagreed, with the other two originally.
184. **A-** It is said that it was proved upon trial that she was informed of a right to vote and had no doubt over her entitlement at the time of voting.
185. **B-** They were charged independently.

Set 38:

186. **C-** Nowhere does it state that all European countries have similar creatures (though certain types can be found in both British Isles and Norway) nor does it state the array of animals is limited to this one nation. Sharing animals and birds does not necessitate sharing geographical features, but it is said a country with forest and moorlands is likely to have a variety of birds and animals, so one can see the link between forests and creatures.
187. **A-** There was a time when the English dreaded wolves and bears, but that indicates the past, or at least does not include the present. Norwegians being superior is not suggested here.
188. **C-** Bears are called destroyers, which is sufficient to conclude they cause damage.
189. **D-** They are ruthlessly hunted by farmers in country districts, but numerous only in the forest tracts in the Far North.
190. **A-** The word fortunately implies that it is good the wolves are no longer central. The children are under no threat, as the threat of wolves belongs to a bygone time, there is no mention of regret that such a time is gone and Norsemen are not demonstrated in the above passage to have respect for Nature, but instead they are said to interfere with it through hunting and driving wolves farther afield from their current homes.

Set 39:

191. **A**- They were seen as the 'friendly or hostile manifestation of some higher powers, demonical and Divine.' The 'manifestation' of the 'Divine' could be interpreted as a religious quality.

192. **C**- A small minority believe that dreams are not the dreamer's own psychical act, meaning the majority believe that they are.

193. **D**- There is incongruity between feelings and images, suggesting a lack of a logical link between the two. Waking thoughts are said to find some dreams repugnant, and though dreams are described as forgotten, that is not necessarily due to them being dull.

194. **A**- It asks can sense be made of each single dream.

195. **C**- A link between the psychical sleeping and waking self is suggested, but not definitively proved. Pre-scientific communities, according to the passage, had a hypothesis that left them with 'no uncertainty'. The 'origin of the dream' remains a question in Freud's writing that has been left without a satisfactory answer. That 'our reminiscences' may 'mutilate' a dream is mentioned, leaving 'C' the only statement with support from the passage.

Set 40:

196. **A**-"Most" requires over half by definition, and "most" of the people living in this area were the descendants of immigrants who moved to the country a "full century ago".

197. **B**- Hall only makes a claim for New England, not the entirety of America, being the descendants of 20,000 immigrants. The 'one million' figure comes from Franklin, not Hall. Less than 80,000 ("under" 80,000) people led to the population boom of one million. One million is over ten times (under) 80,000, so "b" is correct.

198. **C**- It is said to be "distinct" to older aristocracy "of the royal governor's courts". It is not similar to any European aristocracy. There is no specific reference to it not being a system based on lineage.

199. **C**- It says that these were the texts read by the most people, but that does not mean they were the most plentiful – other books may have outnumbered the bibles, even if they received fewer readers.

200. **B**- "A", "c" and "d" are cited in the passage (the journey took 'the better part of the year', it was 'hazardous' and 'expensive'), whereas 'B' is not referenced at all.

Set 41:

201. **D** – Paragraph 2 states that the "main thing" which must be considered when assessing Russia's role is the sapping of the German, Austro-Hungarian and Turkish resources.

202. **5** – America, Italy, Romania and Britain are mentioned at the end of paragraph 1, and France in paragraph 3.

203. **C** – Russia certainly had a greater role when compared to America, Italy, Romania & Britain (paragraph 1). However, there are no details of France's efforts throughout the war.

204. **B** – A & C contributed, but the final catastrophe of the Central Powers was the direct consequence of the offensive of the Allies in 1918 (paragraph 2). Thus, as this was the final catastrophe, this can be concluded to be what won the war.

205. **A** – The final paragraph states that both Russia's and France's efforts were required to stop Germany from the winning the war, which they came very close to.

Set 42:

206. **A** – At the end of paragraph 4, the passage says 'Journalists did some of their finest work and made me proud to be one of them'.

207. **C** – Paragraph 3 states magazines offered perspective, rather than necessarily the facts.

208. **A** – Paragraph 4 states that the internet was responsible for allowing the former audience to write the news.

209. **C** – In paragraph 3 it states 'people gathered around the radio for the immediate news'.

210. **A** – A or B are the 2 most likely answers, as paragraph 2 states about a fallen leader. However, since paragraph 2 only mentions D Roosevelt as being a president (and not John F. Kennedy) this is the best answer.

Set 43:

211. **A** – Paragraph 2 states inhabitants defended their own territory. While paragraph 3 states towns drew up formal treaties with other boroughs, which may include laws of protection, indicating they did, in fact, draw up treaties to defend each other, in addition to defending their own territory.

212. **C** – Paragraph 2 states that inhabitants elected their own rulers and officials in whatever way they themselves chose to adopt.

213. **A** – The first sentence concludes that the 15th century town life is nothing like England today. The last sentence in the second paragraph states there is some comparison with modern America. Therefore, this statement is correct.

214. **C** – The final paragraph states that 'often their authority stretched out over a wide district, and surrounding villages gathered to their markets and obeyed their laws'.

215. **D** – There is mention of war (arming soldiers), of banning trading between boroughs (shutting down privileges of commerce) and treaties in the text.

Set 44:

216. **B** – The second sentence states that male power has risen and thus has not always occurred.

217. **B** – At the end of paragraph 1, it is stated that the Jewish God has attributes of power, thus, D is wrong. There is no mention of the Christian God, therefore, C & A cannot be the answer.

218. **A** – The final sentence talks about how faiths may give an insight into past society.

219. **A** – The first sentence states that 'nowhere is the influence of sex more plainly manifested than in the formulation of religious conceptions and creeds'. Answers B and D could have been right if the question was not *most clearly.*

220. **C** – There is never any mention of the author's own beliefs, there are only discussions of others' beliefs.

Set 45:

221. **True** – The writer says that a "novel with a purpose" constitutes a violation of the unwritten contract tactility existing between writer and reader.

222. **D** – At the end of the first paragraph, the passage states a novel may conduce to delectation (pleasure and delight) during hours of idleness in man. All the other statements the passage states the opposite

223. **C** – The author does not agree with novels which portray the author's views on socialism, religion or divorce laws. However, there is no indication to which the author dislikes the most.

224. **C** – The only plausible answers are C and D (the author argues against A & B). It states that the unwritten contract is a novel should be an 'intellectual artistic luxury', which is earlier described as being able to delectate (to please) – a synonym of amusement. There is no mention of a good novel needing to be fiction.

225. **B** – C and D are beliefs of the author. A novel should instruct the reader is the opinion, while "novel with a purpose" is the answer ('realisation') to this opinion. Therefore, the answer is B rather than A.

Set 46:

226. **C** – While the first paragraph talks of these happy memories, all are asked as questions rather than being presented as fact.

227. **D** – The passage is based on different opinions on what patriotism is, thus, it cannot be deduced that the author can quantify how many American's are patriotic without an agreed definition.

228. **A** – Paragraph 3 states that one proposed definition of patriotism is the training of wholesale murders, i.e. mass killing. There is not mention of depression or workplaces. Children fantasising about having wings to fly to distant lands is not a sufficient reason for D to be the correct answer over A.

229. **C** – While paragraph 1 talks of wonderful tales of great deeds and conquests, this is insinuated to be of the past. The author then in paragraph 2 says that the stories mothers tell today are those of sorrow, tears, and grief.

230. **D** – This is the only answer where the author specifically says children wonder. The others could be inferred but are not as factually correct.

Set 47:

231. **C** – The start of paragraph 2 states that a good man can receive a gift well.
232. **A** – In paragraph 1, it states "we wish to be self-sustained" when referring to some men's opinions.
233. **B** – In paragraph 2, the author states that he can rejoice (i.e. like) to receive gifts as well as sometimes grieving upon the presentation of a gift. He later goes on to say he is ashamed of being pleased by gifts. However despite this, for certain periods, he does sometimes like gifts.
234. **D** – In paragraph 1, it states "we can receive anything from love, for that is a way of receiving it from ourselves".
235. **C** – Paragraph 2 states "if the gift pleases me overmuch, then I should be ashamed that the donor should read my heart, and see that I love his commodity, and not him." I.e. in this situation, the author feels bad that he does not appreciate the unique effort and insight of the giver as much as the physical gift. Answer B could be inferred, however, the answer C is stated. A is not true and so D cannot be the answer.

Set 48:

236. **A** – False. The final sentence states that mathematics was a science which Goethe was imperfectly acquainted with.
237. **C** – Having writers speak about a paper indicates this is the most popular (the first sentence). But it is specific parts of this paper (prismatic spectrum and refraction) which were mostly spoke about. Therefore, C rather than A is correct.
238. **A** – The translator was aware of the opposition which the theoretical views alluded to conflict with and wanted to first select the experiments which showed evidence to support Goethe's view. It says this was incompatible with the author's view.
239. **C** – "Doctrine of Colours" and two short essays entitled "Contributions to Optics" are mentioned.
240. **A** – This is stated in the first sentence. While his views are different from the received theories of Newton, it does not say it disproves Newton's theories.

Set 49:

241. **True** – In paragraph 2 the passage says "gradually" (but surely) the thoughts of money takes over the thoughts of a girl so that income is the most important thing – therefore, based on the information solely in the passage, this is true.
242. **False** – In paragraph one, the passage states that this is the view of stereotypical Philistine's and therefore not the author's view.
243. **Can't tell** – Nowhere does the passage mention Nora's opinion of wanting a child (be it with anyone).
244. **Can't tell** – The first paragraph states every twelfth marriage ends in divorce, but nowhere does it say what population of people this statistic was calculated from. In paragraph 6, it talks of American life but this does not necessarily mean the statistic is from America.
245. **False** – In the second paragraph, it states that over time, a married woman's mind strays from these thoughts.

Set 50:

246. **False** – In paragraph 1, it states a series of temporary measures were initiated (somewhere in the 1960s/70s). However, in paragraph 2 it says the Terrorism Act 2000 was passed as a definite measure following twenty years of temporary measures, therefore, the temporary measures in Ulster could not have been implemented in the 60s, as this would be over thirty years.
247. **Can't tell** – Nothing stated anywhere in the text of legislation actively replacing it.
248. **Can't tell** – Paragraph 2 would suggest so, but there is no definitive description of why the acts were passed like there is in paragraph 1.
249. **True** – Paragraph 3 states that the ATCSA marked a more firm move towards the 'management of anticipatory risk' which was to characterize the counter-terrorism legislation of the 21st century.
250. **True** – The second paragraph states that the 2000 Terrorism Act was passed "following twenty years of temporary measures."

Decision Making Answers

Question	Answer	Question	Answer	Question	Answer	Question	Answer
1	D	26	D	51	C	76	B
2	B	27	B & D	52	B	77	A
3	B	28	A & D	53	D	78	C
4	B & C	29	A	54	A	79	A
5	D	30	C	55	A	80	A
6	C	31	B	56	D	81	A
7	C	32	A & C	57	C	82	B
8	B	33	B & D	58	A	83	C
9	C & E	34	C	59	A	84	All false
10	B	35	D	60	D	85	B, C & D
11	C	36	D	61	C	86	C
12	B	37	C	62	C	87	D
13	B	38	B	63	B	88	A
14	B	39	A	64	D	89	B & D
15	C	40	A	65	C	90	B
16	C	41	C	66	D	91	B
17	A	42	A	67	C	92	C
18	D	43	D	68	D	93	A
19	E	44	D	69	A	94	B
20	C	45	A	70	D	95	B
21	D	46	B	71	B	96	C
22	D	47	A	72	D	97	A
23	C	48	B	73	D		
24	D	49	D	74	D		
25	B	50	C	75	D		

Question 1: D

The easiest thing to do is draw the relative positions. We know Harrington is north of Westside and Pilbury. We know that Twotown is between Pilbury and Westside. Crewville is south of Twotown, Westside and Harrington but we do not know but its location relative to Pilbury.

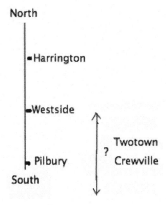

Question 2: B

By making a grid and filling in the relevant information the days Dr James works can be deduced:

	Sunday	Monday	Tuesday	Wednesday	Thursday	Friday	Saturday
Dr Evans	X	√	X	X	√	√	√
Dr James	X	√	√	√	√	X	√
Dr Luca	X	X	√	√	X	√	√

- ➢ No one works Sunday.
- ➢ All work Saturday.
- ➢ Dr Evans works Mondays and Fridays.
- ➢ Dr Luca cannot work Monday or Thursday.
- ➢ So, Dr James works Monday.
- ➢ And, Dr Evans and Dr James must work Thursday.
- ➢ Dr Evans cannot work 4 days consecutively so he cannot work Wednesday.
- ➢ Which means Dr James and Luca must work Wednesday.
- ➢ (mentioned earlier in the question) Dr Evans only works 4 days, so cannot work Tuesday.
- ➢ Which means Dr James and Luca work Tuesday.
- ➢ Dr James cannot work 5 days consecutively so cannot work Friday.
- ➢ Which means Dr Luca must work Friday.

Question 3: B

All thieves are criminals. So the circle must be fully inside the square, we are told judges cannot be criminals so the star must be completely separate from the other two.

Question 4: B and C

Using the information to make a diagram:

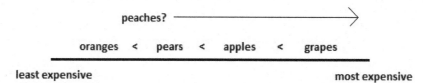

Hence **A** is incorrect. **D** and **E** may be true but we do not have enough information to say for sure. **B** is correct as we know peaches are more expensive than oranges but not about their price relative to pears. Equally we know **C** to be true as grapes are more expensive than apples so they must be more expensive than pears.

Question 5: D

Using Bella's statements, as she must contradicted herself with her two statements, as one of them must be true, we know that it was definitely either Charlotte or Edward. Looking to the other statements, e.g. Darcy's we know that it was either Charlotte or Bella, as only one of the two statements saying it was both of them can have been a lie. Hence it must have been Charlotte.

Question 6: C

Work through each statement and the true figures.

A. Overlap of pain and flu-like symptoms must be at least 4% (56+48-100). 4% of 150: 0.04 x 150=6
B. 30% high blood pressure and 20% diabetes, so max percentage with both must be 20%. 20% of 150: 0.2*150 = 30
C. Total number of patients – patients with flu-like symptoms – patients with high blood pressure. Assume different populations to get max number without either. 150 – (0.56 x 150) – (0.3 x 150) = 21
D. This is an obvious trap that you might fall into if you added up the percentages and noted that the total was >100%. However, this isn't a problem as patients can discussed two problems.

Question 7: C
We know that Charles is born in 2002, therefore in 2010 he must be 8. There are 3 years between Charles and Adam, and Charles is the middle grandchild. As Bertie is older than Adam, Adam must be younger than Charles so Adam must be 5 in 2010. In 2010, if Adam is 5, Bertie must be 10 (states he is double the age of Adam). The question asks for ages in 2015: Adam = 10, Bertie = 15, Charles = 13

Question 8: B
In this question it is worth remembering it will take more people a shorter amount of time.
Work out how many man hours it takes to build the house. Days x hours x builders
12 x 7 x 4 = 336 hours
Work out how many hours it will take the 7man workforce: 336/7 = 48 hours
Convert to 8 hour days: 48/8 = 6 days

Question 9: C & E
The easiest way to work this out is using a table. With the information we know:

1st		Madeira
2nd		
3rd	Jaya	
4th		

Ellen made carrot cake and it was not last. It now cannot be 1st or 3rd as these places are taken so it must be second:

1st		Madeira
2nd	Ellen	Carrot cake
3rd	Jaya	
4th		

Aleena's was better than the tiramisu, so she can't have come last, therefore Aleena must have placed first

1st	Aleena	Madeira
2nd	Ellen	Carrot cake
3rd	Jaya	
4th		

And the girl who made the Victoria sponge was better than Veronica:

1st	Aleena	Madeira
2nd	Ellen	Carrot cake
3rd	Jaya	Victoria Sponge
4th	Veronica	Tiramisu

Question 10: B
After the first round; he knocks off 8 bottles to leave 8 left on the shelf. He then puts back 4 bottles. There are therefore 12 left on the shelf. After the second round, he has hit 3 bottles and damages 6 bottles in total, and an additional 2 at the end. He then puts up 2 new bottles to leave 12 − 8 + 2 = 6 bottles left on the shelf. After the final round, John knocks off 3 bottles from the shelf to leave 3 bottles standing.

Question 11: C
Remember that pick up and drop off stops may be the same stop, therefore the minimum number of stops the bus had to make was 7. This would take 7 x 1.5 = 10.5 minutes.
Therefore the total journey time = 24 + 10.5 = 34.5 minutes.

Question 12: B

The time could be 21:25, if first 2 digits were reversed by the glass of water (21 would be reversed to give 15). **A** cannot be the answer, because this would involve altering the last 2 digits, and we can see that 25 on a digital clock, when reversed simply gives 25 (the 2 on the left becomes a 5 on the right, and the 5 on the right becomes a 2 on the left). **C** cannot be the answer, as this involves reversing the middle 2 digits. As with the right two digits, the middle 2 digits of 2:5 would simply reverse to give itself, 2:5. **D** could be the time if the 2nd and 4th digits were reversed, as they would both become 2's. However, the question says that 2 *adjacent* digits are reversed, meaning that the 2nd and 4th digits cannot be reversed as required here. **E** is not possible as it would require all four numbers to be reversed. Thus, the answer is **B**.

Question 13: B

To answer this, we simply calculate how much total room in the pan will be taken up by the food for each guest:
- 2 rashers of bacon, giving a total of 14% of the available space.
- 4 sausages, taking up a total of 12% of the available space.
- 1 egg takes up 12% of the available space.

Adding these figures together, we see that each guest's food takes up a total of 38% of the available space.

Thus, Ryan can only cook for 2 guests at once, since 38% multiplied by 3 is 114%, and we cannot use up more than 100% of the available space in the pan.

Question 14: B

The trains come into the station together every 40 minutes, as the lowest common multiple of 2, 5 and 8 is 40. Hence, if the last time trains came together was 15 minutes ago, the next time will be in 25 minutes.

Question 15: C

Tiles can be added at either end of the 3 lines of 2 tiles horizontally or at either end of the 2 lines of 2 tiles vertically. This is a total of 10, but in two cases these positions are the same (at the bottom of the left hand vertical line and the top of the right hand vertical line). So the answer is $10 - 2 = 8$.

Question 16: C

Georgia is shorter than her Mum and Dad, and each of her siblings is at least as tall as Mum (and we know Mum is shorter than Dad because Ellie is between the two), so we know Georgia is the shortest. We know that Ellie, Tom and Dad are all taller than Mum, so Mum is second shortest. Ellie is shorter than Dad and Tom is taller than Dad, so we can work out that Ellie must be third shortest.

Question 17: A

Danielle must be sat next to Caitlin. Bella must be sat next to the teaching assistant. Hence these two pairs must sit in different rows. One pair must be sat at the front with Ashley, and the other must be sat at the back with Emily. Since the teaching assistant has to sit on the left, this must mean that Bella is sat in the middle seat and either Ashley or Emily (depending on which row they are in) is sat in the right hand seat. However, Bella cannot sit next to Emily, so this means Bella and the teaching assistant must be in the front row. So Ashley must be sat in the front right seat.

Question 18: D

We can see from the fact that all the possible answers end "AME" that the letters "AME" must be translated to the last 3 letters of the coded word, "JVN", under the code. J is the 10th letter of the alphabet so it is 9 letters on from A (V is the 21st letter of the alphabet and M is the 13th, and N is the 14th letter of the alphabet and E is the 5th, therefore these pairs are also 9 letters apart). Therefore P is the code for the letter 9 letters before it in the alphabet. P is the 16th letter of the alphabet, therefore it is the code for the 7th letter of the alphabet, G. Therefore from these solutions the only possibility for the original word is GAME.

Question 19: E

To find out whether many of these statements are true it is necessary to work out the departure and arrival times, and journey time, for each girl.

Lauren departs at 2:30pm and arrives at 4pm, therefore her journey takes 1.5 hours

Chloe departs at 1:30pm and her journey takes 1 hour longer than 1.5 hours (Lauren's journey), therefore her journey takes 2.5 hours and she arrives at 4pm

Amy arrives at 4:15pm and her journey takes 2 times 1.5 hours (Lauren's journey), therefore her journey takes 3 hours and she departs at 1:15pm.

Looking at each statement, the only one which is definitely true is **E**: Amy departs at 1:15pm and Chloe departs at 1:30pm therefore Amy departed before Chloe.

D *may* be true, but nothing in the question shows it is *definitely* true, so it can be safely ignored.

Question 20: C

For the total score to be odd, there must be either three odd or one odd and two even scores obtained. Since the solitary odd score could be either the first, second or third throw there are four possible outcomes that result in an odd total score. Additionally, there are the same number of possibilities giving an even score (either all three even or two odd and one even scores obtained), and the chance of throwing odd or even with any given dart is equal. Therefore, there is an equal probability of three darts totalling to an odd score as to an even score, and so the chance of an odd score is ½.

Question 21: D

A. Not appropriate as it is highly judgemental towards the abilities of nurses. Nurses already play an important role in care delivery and in many cases already make important decisions. The NHS would not be able to deliver care efficiently without the input of specialty nurses.
B. Worse version of statement a). Same reasons why this is inappropriate.
C. In general, not completely wrong, but the fact that it places the ability of nurses over these of doctors without limitation is the flaw of this argument. Nurses and doctors work together to deliver care with different areas of responsibility.
D. Correct answer. True and less restrictive than the version in c). Nurses input　valuable and it would definitely be helpful to give nurses more freedom for decision making.

Question 22: D

a) is wrong because the text specifically states that he sees many squirrels. B) is wrong because the birds fly off as soon as James approaches. C) is an assumption not backed by the text resource and therefore must be wrong.

Question 23: C

A. Incorrect. This is not about being all-knowing and infallible. It is about the assumption of do no harm.
B. Training is not mentioned in the quote at all, therefore this answers is incorrect. It can be further assumed that for a physician to be able to tell the present and future, he must have undergone some degree of training.
C. Correct. Non-maleficence means do no harm.
D. In some cases techniques of the past are valuable and still used today, in some cases they are not. In any case this quote is not about techniques, but about the moral background of actions.

Question 24: D

A. Incorrect. This would solve the problem but ignore all ethical basis we use for medical decision making.
B. Incorrect. This would cause a significant impact to the everyday life of the families of the patients making it a highly unethical decision.
C. Incorrect. The entire idea of the NHS is based on the notion that there is no discrimination between the rich and the poor. It is for that reason unacceptable to charge patients for extended hospital stays unless they get different care.
D. Correct. All other answers are incorrect per descriptions above.

Question 25: B

The relationship of goats and pigs to cows is 4:1 meaning that in every 5 animals, 1 has to be a cow. The value of the individual animals is a distraction that is not needed for the calculation of this question. 5 = 40

Question 26: D

A. Incorrect. Young apes mingle with any group as it says in the text.
B. Incorrect. The widest range of social interactions is displayed by the youngest apes which mingle freely with any group of monkeys.
C. Incorrect. Has nothing to do with the actual question.
D. Correct. There seems to be a separation between adolescent male and female apes as their preferred groups of social interaction are separated. It says specifically in the text that adolescent female apes mingle exclusively with young and older female apes.

Question 27: B & D

A. No. This conclusion cannot be drawn as no information is given regarding the friend. He/she could simply be of smaller statue. Also, it is unknown what mount of cooking equipment the two are taking.
B. Yes. They both carry equipment that is important for them having a good time.
C. No. Similar to a) the simple fact that there is a chance that Steve's load may be heavier than his friends does not provide enough information to deduct his friend to be female.
D. Yes. Due to the amount of kit the two are carrying, most importantly the tent and sleeping equipment suggests that the two want to stay wherever they are going at least for one night.

Question 28: A & D

Order of seats from left to right: Sophie – Maria – Anna – Louise – Jenny

A. True. The only unallocated seat is between Sophie and Anna, therefore Maria must sit on the second seat from the left.
B. False. Sophie sits at the edge with regards to the row occupied by the 5 girls, the text never suggests that she mat sit at the edge of the hall. As a matter of fact, since the text specifically states that they sit in front of the stage, it is unlikely that Sophie sits at the edge of the concert hall.
C. False. Jenny sits at the far right.
D. True. Louise sits between Jenny and Anna.

Question 29: A

A. True. Sven plants salad, carrots and courgettes, making it three. Tomatoes are strictly speaking fruits, so they are not counted.
B. False. Carrots are orange.
C. False. The text does not deliver any data regarding the prevalence of tomato allergies. All we know is that Sven has a tomato allergy.
D. False. Conclusion a) is true.

Question 30: C

A. False. Rome is the most Southern city.
B. False. According to the travel organisation, Amsterdam must be further South than Berlin.
C. True. They will start in the UK, move on to German, then the Netherlands, then France, then Italy.
D. False, c) is true.

Question 31: B

A. False. The text clearly states that computer technology firms are heavily involved in military contracts.
B. True. The only food producing industry branch that is involved in military contracts according to the text is the beef industry.
C. False. This conclusion cannot reasonably be made as the data in the text passage is insufficient.
D. False. This uses information not mentioned in the text source at all.

Question 32: A & D
Order of houses: Austin – Peter – Steve – David – Mark
A.   True. Steve lives in the middle of the row of houses.
B.   False. There is no information about the size of the town.
C.   True. Peter must live between Steve and Austin as this is the only available house that is not allocated by the description.
D.   False. David lives between Steve and Mark.

Question 33: B & D
A.   False. Whilst it is true that baldness can affect women as well, the study only focuses on men.
B.   True. Both anabolic steroid consumption as well as type 2 diabetes can be linked to life style choices, therefore there is a connection between hair loss and life style changes.
C.   False. The majority of cases are due to a genetic predisposition and independent of external factors. It can therefore be assumed that shampoo will not help with hair loss.
D.   True. The link is directly mentioned in the text.

Question 34: C
A.   False. The forest also contains fruit trees.
B.   False. False. There are needle trees, oaks, beeches and fruit trees..
C.   True. Christmas trees are needle tress which represent the majority of trees in the forest.
D.   False. We can't answer that question as no actual size of the forest is given.

Question 35: D
A.   True. Since they spend 50% of their life in the ocean, the other 50% must be on land since they are flightless.
B.   True. Penguins life in large groups making them herd animals.
C.   True. Says so specifically in the text.
D.   False. Penguins live in various climate regions as it says in the text.

Question 36: D
A.   False. Fossil fuel technology has constantly been improved in efficiency and has not been overcome to this day.
B.   False. Import of fossil fuels such as coal and oil is expensive, but not expensive enough to make it economically unsound as it forms an essential part of most industrial productions steps.
C.   False. Fossil fuels are limited.
D.   True. Due to our huge energy demand, an immediate switch to non-fossil fuels is difficult. Also, there are less developed countries where fossil fuels represent the primary source of energy sue to technological constraints regarding energy generation.

Question 37: C
A.   False. There are some parts of the population that are not allowed to vote as they have not yet reached the legal voting age.
B.   False. This is only in part responsible for the decision not to vote.
C.   True. Voters must be of a certain age and also be allowed to vote based on other criteria.
D.   False.

Question 38: B
A.   False. Older parents are more likely to have genetically abnormal children.
B.   True. Due to the accumulative mutagenic effect the risk of genetic alteration due to environmental mutagens is lower in younger parents.
C.   False. Sperm cells are constantly regenerated. It is egg cells that are present at birth.
D.   False. Answer b is correct.

Question 39: A

A.   True. Peter is 3 seconds faster than Felix, who is 5 seconds faster than Lucas.

B.   False, Dorian finishes second, and according to the text the top 3 places are separated by 1 second.

C.   False. Dorian finishes second.

D.   False. Answer a) is correct.

Question 40: A

The sum of each circle must represent the amount of people with the specific preferences.

Gummi bears: 5, of which 2 like chocolate and 2 like biscuits.

Chocolate: 6, of which 2 also like gummi bears and 2 like biscuits. 2 only like chocolate.

Biscuits: 3, of which 1 also likes gummy bears and 2 also like chocolate.

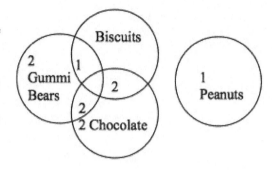

Question 41: C

As the all the batches have the same value and the company has sold less candy. The previous year the company sold 38 322 tons of candy and this year they only sold 32 500 tons of candy.

Calculations:

Product 1: 10 000 = 110% of last year's sales => last year: 9091

Product 2: 500 = 5% of last year's sales => last year: 10000

Product 3 12 000 = 130% of last year's sales => last year: 9231

Product 4 and 5 are unchanged.

Question 42: A

The assumption that the surgeon only does private work is not supported by the resource as he only lists his 4 main procedures. It should not be assumed that this is the only work he does.

Question 43: D

Since there is a 50% chance of one of the two being a smoker, there is a 25% chance of both f of them being smokers.

Question 44: D

The question does not address issues such as income or relationships of power. Whilst indirectly, answers a) and c) may well be true, there is no factual information in the question to back up that claim.

Question 45: A

50% of cases due to respiratory problems. 25% of cases due to cardiovascular causes, 15% of cases due to abdominal disease, 5% due to traffic accidents, 5% due to work accidents.

Question 46: B

65% of 15% equates to 10%. 15% does not represent a large proportion, therefore answer a) must be incorrect. Nuts are not the most common source of allergies according to the text, in general food intolerances represent the most common cause for allergies, therefore c) is wrong. The question is specifically about children and for this reason, answer d) is incorrect too.

Question 47: A

The fact that chimpanzees continuously use tools shows that they understand their effect, this makes b) true and therefore the wrong answer. With regards to answer c), using tools in itself represents a form of environmental manipulation making c) true and therefore the incorrect answer for this question.

Question 48: B

Carl weighs 50kg, Luke weighs 45kg, Ben and Alex weigh 40kg and Peter weighs 35kg. In total, they weigh 210kg, meaning the house structure will weigh 15kg according to their calculation.

Question 49: D

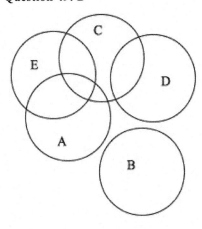

Question 50: C

Ants do not compete amongst individuals for resources but amongst colonies of the same species, therefore a) is incorrect. For similar reasons, b) must be incorrect as different species may have different food preferences meaning that there is no interspecies competition. D) is incorrect as the text specifically states that some species of ants live of plant material, whilst others live off animal material.

Question 51: C

Many pesticides carry health risks for the consumer and are present in the food we consume every day, which carries significant health risks.

Question 52: B

False – True – False – False.

Question 53: D

Answer a) is true as I mentioned in the text. Answer b) is true as farming formed the basis to feed a population restricted to one location. C) is true as crops need time to grow and harvest which is not conducive to a nomadic lifestyle.

Question 54: A

All of the conclusions are true. Since the goal of school is the intellectual development of children, increasing exercise may help improve performance. A sedentary life style ins marked by a lack of exercise, therefore this may contribute to poorer school performance. C) is basically the original statement, but rephrased.

Question 55: A

The mere fact that these plants stand in a similar constellation to their sun, as Earth does to our sun is only one part of the equation. This is relevant because it increases the likelihood of liquid water anywhere near the surface of the plant, which, next to air is one of the most essential points. The other assumptions are largely irrelevant for questions of habitability.

Question 56: D

The fact that species C lives of berries when available but then falls back to seeds or insects means that they can find food in any season. Answer a) Is incorrect as only species C eats berries, answer b) is incorrect as there is no mention of their beaks in the text, answer c) is incorrect as Species A will also eat sunflower seeds.

Question 57 C

Whether health-care professionals are role-models for society or not may be debatable, but they represent important points of contact between the public and the healthcare system giving them certain responsibilities, therefore a) is incorrect. B) is incorrect as it is not related to the question. D) is true, but irrelevant to the question and therefore is false.

Question 58: A.

A.   Correct as per the final sentence of the passage
B.   incorrect as per text
C.   Incorrect as the text is about young men and women are not mentioned anywhere.
D.   Incorrect as option (a) is correct.

Question 59: A

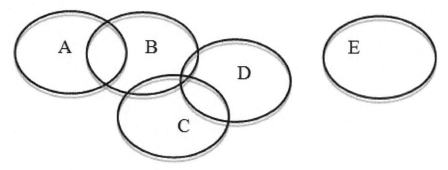

Population A: Applicants with excel skills
Population B: Applicants with excel and Photoshop skills
Population C: Applicants with excel, Photoshop and language skills
Population D: Applicants with excel, Photoshop, language and communication skills.
Population E: Applicants with only one language but experience.

Question 60: D

The overall order is: Tom - Richard - Anthony - Steve - Tony
The text describes that Richard finished joint bottom in section C, therefore he did not finish well in the section.

Question 61: C

A.   Incorrect. The new cigarette blend releases twice as much CO.
B.   False. The new cigarette contains less tar.
C.   True. The new cigarette contains twice as much nicotine than the competition meaning one has to smoke two competitor cigarettes for every on to reach the same level of nicotine.
D.   Incorrect. In all aspects mentioned, except for the tar, the new blend is worse than the competitions product.

Question 62: C

Elmsworth to Eastwich:	30 minutes
Waiting time at Eastwich:	5 minutes
Eastwich to Northtown:	45 minutes
Waiting time at Northtown:	5 minutes
Northtown to Southwarf:	30 minutes
Stop at Northtown station:	5 minutes
Total:	120 minutes

Question 63: B

Team A: 7 points.
Team B: 3 points.
Team C: 4 points.
Team D: 12 points.
Team E: 1 point.

A.   False. Since team D is unbeaten and each team plays each opponent only once, they achieve the most points.
B.   True. Since team B finishes with 3 points they must draw three times, with team A, C and E.
C.   False. Team C will collect fewer points than team D.
D.   Irrelevant therefore false.

Question 64: D
A. Incorrect. Whilst the war is over. Resource conservation makes sense none the less.
B. Incorrect. Not a strong enough argument.
C. Possibly true, but not a big problem.
D. Correct. An issue that demonstrates both urgency as well necessity.

Question 65: C
The progression of the race is as follows: Start -15km- Chris -10km- Philip -6km- Anne -5km- Tara -4km- Peter -2km- finish. Therefore the distances between stops progressively reduce through the race

Question 66: D
Whilst all other options are valid, the last is the by far strongest issue when it comes to quality research.

Question 67: C
Stephanie – Sophie – Patricia – Tina – Annette

Question 68: D
A. The text states that there is a high degree of variability in decision making.
B. The text states that there is no regard for social background.
C. Similar reason as statement b.

Question 69: A
Cambridge – Imperial – Sheffield – Southampton – Cardiff

Question 70: D
A. Very superficial, delivers no actual argumentative value.
B. True, but not the strongest argument.
C. Similar to a). Also leads in a different direction than the argument itself.
D. True and the most basic and most central argument in support of the problem as described by the statement.

Question 71: B
A. False. The passage is about Tim, not Tom.
B. True. The passage says this is his favourite method.
C. Possibly true, but no comparison of strategies is made.
D. False.

Question 72: D
A. False. This statement only investigates small fish.
B. False. Water temperature is never mentioned.
C. False. The statement directly mentions that worse eyesight improves individual survival chances.
D. True. In cases of impaired visual function such as through dirt in murky water or through worse eye sight of the predator, the individual seems to be better protected.

Question 73: D
A. Incorrect. Since 50% of people consider healthcare to be the number one issue, there is clearly a problem with heath care provision.
B. Incorrect. Only because 10% consider defense the most important issue does not mean the population feels very secure. It just means they consider other points more important.
C. False. Irrelevant to the question.
D. Correct. Healthcare = 50%, education = 25%, infrastructure = 15%, defense = 10%

Question 74: D

A. Correct. Health care accounts for 22% whereas public cleanliness is 16%.
B. Correct. Housing accounts for 24% whereas healthcare is at 22%
C. Correct at 11%.
D. Incorrect. Since education accounts for 28% and housing for 5% less, this makes it 23%.

Question 75: D

A. Incorrect. The text specifically states that this happens.
B. Incorrect. Whilst this might be true in reality, it is not supported by the text.
C. Incorrect. According to the text, males attract females with their mating calls.
D. Correct.

Question 76: B

A. Incorrect. The groups are led by females.
B. Correct. The purpose of the group is to protect the young.
C. Incorrect. The groups are led by an individual according to the text.
D. Incorrect.

Question 77: A

A. Correct.
B. Incorrect. They eat two colours of fruit.
C. Incorrect. We only know that monkeys don't like to eat them.
D. Incorrect as irrelevant.

Question 78: C

A. Incorrect. Whilst without members, there can be no group, this presents little with regards to support for the statement.
B. Irrelevant – not mentioned in question.
C. Correct a strong and plausible rationale.
D. Incorrect.

Question 79: A

A. Correct.
B. Incorrect. Since C and B are the same, they both sold 2000 units.
C. Incorrect. The text does not address price.
D. False. Set D is the second to last most commonly sold handset.

Question 80: A

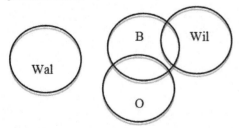

Walnut tree cannot grow around other trees due to the poison. The other trees both grow with Beech but Oak and Willow do not grow well together.

Question 81: A

A. Correct. 80% of 5 represent 4.
B. Incorrect. There are no vegan dishes in the order.
C. Incorrect. Whilst this might be true, there is no evidence to support this.
D. Incorrect.

Question 82: B
A. False. Since only gold medals are considered, Group A falls behind group B
B. True, it must have more gold medals to rank in top position
C. False as per text.
D. False.

Question 83: C
A. Incorrect. The statement is true.
B. Incorrect as the statement is true.
C. Correct as climate change causes increased temperatures in the normal breeding grounds rather than in the South.
D. Incorrect as the statement is true.

Question 84: All false
A. False – the passage describes coloured clothes
B. False – no red clothes are described
C. False – he owns sports clothes
D. False – some are but we cannot conclude that most are

Question 85: B, C & D
Peter - Eugene – John – David – Anthony.
Therefore (a) = false, (b) = true, (c) = true, (d) = true.

Question 86: C
A. True, but not the strongest argument.
B. False.
C. True, strongest argument.
D. False, as it replaces one unfair restriction with another.

Question 87: D
A. False. Per the text bears do not hunt.
B. False. Bears prefer berries over any other food source.
C. False. Bears attack if their cubs are in danger.
D. True.

Question 88: A

Because some ants are exclusively dedicated to construction work, and male ants are exclusively for reproduction, the correct option shows both of these independently.

Question 89: B & D
A. False – there are cherry trees.
B. True – all the described plants are trees.
C. False – her husband plants cherry trees.
D. True – raspberry bushes are all she planted; the other plants were either there before or planted by her husband

Question 90: B
A. Partially true, certainly not the strongest argument.
B. Correct. The term "unhealthy foods" is not well enough defined and can vary greatly.
C. False. Obesity is not mentioned in the argument.
D. False.

Question 91: B
A.   Incorrect. This is not an actual quantitative value.
B.   Correct since the text states that using nails will make the wardrobe more likely to collapse.
C.   Incorrect. The text specifically states, that Walnut is darker than pine.
D.   Incorrect. Peter does not play a role in the question.

Question 92: C
A.   Incorrect.
B.   Incorrect. At no point is the material named specifically.
C.   True. As material C contains a significant proportion of the expensive material B, it must be purchased making material C expensive.
D.   Incorrect. They must interact in order to produce material C which has different properties than its basic components.

Question 93: A
A.   Correct.
B.   False. Company B makes $50,000, Company C makes $90,000
C.   False. The text specifically states, that it is the smallest company.
D.   False. The text only says that this company has the smallest income which does not mean that it is going bankrupt.

Question 94: B

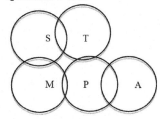

This is quite a complex one, so it may help to draw a 5x5 table to compare everyone's relationships. Identify the people who get on well to work out which people should overlap on the Venn diagram. Notice that Sarah gets on with Tina and Marylin only. Peter gets on with everyone except Sarah. These two pieces of information are enough to make option B the clear choice.

Question 95: B
A.   Incorrect and irrelevant. Not addressed in the statement.
B.   Correct. As population continues to grow newspaper numbers reduce.
C.   False. This is a matter of opinion.
D.   False.

Question 96: C
A.   Incorrect. The same number of children want to be actors as well as politicians, 20 children.
B.   Incorrect. Since 80 children don't know what they want to be yet, this is the majority.
C.   True.
D.   False. Since the same number of children want to be actors and politicians (20), more children want to be teachers than politicians.

Question 97: A
A.   True – as per text
B.   False – neither score
C.   False – information not in passage
D.   False – information not in passage, we only know he scores fewer goals

Quantitative Reasoning Answers

Q	A	Q	A	Q	A	Q	A	Q	A	Q	A
1	C	51	A	101	A	151	C	201	E	251	C
2	D	52	E	102	E	152	B	202	B	252	B
3	D	53	C	103	A	153	C	203	C	253	A
4	D	54	E	104	C	154	B	204	D	254	C
5	C	55	C	105	B	155	B	205	C	255	C
6	B	56	A	106	C	156	D	206	B	256	D
7	A	57	D	107	C	157	D	207	C	257	C
8	A	58	D	108	B	158	C	208	A	258	A
9	B	59	C	109	B	159	B	209	C	259	D
10	A	60	D	110	C	160	A	210	C	260	D
11	D	61	C	111	C	161	B	211	B	261	A
12	B	62	A	112	C	162	B	212	C	262	D
13	A	63	E	113	C	163	A	213	D	263	A
14	B	64	E	114	A	164	B	214	B	264	A
15	B	65	D	115	A	165	A	215	D	265	C
16	D	66	B	116	B	166	D	216	A	266	D
17	B	67	C	117	C	167	A	217	B	267	A
18	D	68	D	118	A	168	B	218	C	268	D
19	B	69	E	119	B	169	B	219	A	269	A
20	B	70	B	120	D	170	C	220	B	270	D
21	B	71	C	121	B	171	B	221	C	271	E
22	B	72	E	122	C	172	C	222	B	272	A
23	C	73	D	123	C	173	B	223	E	273	D
24	B	74	A	124	A	174	B	224	B	274	B
25	B	75	E	125	B	175	D	225	D	275	E
26	B	76	C	126	A	176	D	226	C	276	D
27	D	77	E	127	B	177	B	227	E	277	B
28	B	78	D	128	C	178	D	228	C	278	E
29	B	79	D	129	C	179	D	229	B	279	A
30	B	80	D	130	D	180	C	230	D	280	A
31	C	81	A	131	C	181	D	231	D	281	C
32	B	82	C	132	A	182	A	232	C	282	C
33	B	83	E	133	C	183	D	233	E	283	E
34	D	84	B	134	B	184	C	234	D	284	C
35	B	85	D	135	C	185	B	235	C	285	A
36	D	86	E	136	B	186	A	236	B	286	D
37	C	87	D	137	C	187	C	237	C	287	C
38	D	88	B	138	A	188	D	238	C	288	E
39	E	89	B	139	B	189	A	239	B	289	E
40	C	90	B	140	B	190	C	240	D	290	C
41	C	91	C	141	D	191	B	241	A	291	A
42	A	92	D	142	C	192	D	242	B	292	D
43	E	93	C	143	B	193	B	243	E	293	E
44	A	94	E	144	D	194	C	244	B	294	A
45	E	95	E	145	B	195	A	245	D	295	D
46	C	96	B	146	C	196	D	246	A	296	C
47	A	97	D	147	B	197	D	247	B	297	D
48	D	98	C	148	B	198	C	248	B	298	B
49	D	99	D	149	B	199	C	249	B	299	D
50	C	100	A	150	D	200	A	250	C	300	E

SET 1

Question 1: C
We can work out that the tax rates must fit in the following equation:
($50 x Food tax rate) + ($30 x Clothes tax rate) + $80 = $88. Only the tax rates in Casova fit correctly in this equation.

Question 2: D
To answer this question, we calculate how much the supplier will make for selling the items, by considering the tax rate in each state, and deducting it from the price accordingly.

Thus, in Bolovia, each year the supplier makes 250 x ($40/1.20 + $40/1.15 + $40/1.10 + $40/1.15) = $34,812 a year. In Asteria, each year the supplier makes 250 x ($40/1.10 + $40/1.15 + $40/1.10 + $40/1.15) = $35,572. (Note that in the case of an item being applicable to 2 tax rates, the higher rate will be charged. Thus, in Bolovia, imported clothes will be charged at the clothes tax rate of 15%, since this is higher than the imports rate.)
Thus, by moving to Asteria, the supplier will make $760 more each year. Therefore it will take 26.3 years to recover the purchase cost of $20,000.

Question 3: D
If John spends $88, he will spend £12 on tax. Thus, the tax rate is 12/88 = 13.6%.
If John shops in Asteria, the maximum tax rate he would have to pay is 10%; at Casova it would be 10%. If he spends at least $50 on food in Derivia, he pays no tax on it. Thus, he can spend a maximum of $38 on imported goods (at a maximum tax rate of 15%). This equates to a tax of $5.70 (not $12). Finally, if John spends $10 on imported goods in Bolovia – he would pay $0.50 in tax. Thus, he can spend up to $78 on clothes taxed at 15%. The tax on the clothes is therefore $11.70, giving $12.20 tax in total as the maximum. Since he pays $12 dollars tax, he shops in Bolovia.

Question 4: D
The sum of the basic prices is 100+30+10+100 = $240. Now the highest tax rate on the board is 20% (for imports to Asteria), thus the maximum tax is $240 x 1.20 = $288. However, this is impossible to attain (since if we bought everything in Asteria, the ham would be cheaper, as it is not imported and would only be taxed at the food rate). Therefore no option allows the overall price to be as high as $288, so this is the answer. Answer A) is possible if all products were bought in the state they are produced in. Answer C) is the correct answer if all products were bought in Asteria (and accounting for the reduced tax rate for the ham, which is not an import). Answer B) is possible if the ham was bought in Asteria, the caviar and orange juice were bought in Casova and the dress was bought in Bolovia.

SET 2

Question 5: C
Firstly, find the pressure it can withstand in Pascals: 200 pounds per square inch x 7,000 Pascals per pound per square inch = 1.4 million Pascals.
Then divide this by 1,000 Pa to get the depth the probe can withstand (we can see from the question that the pressure increases by 1,000 Pa for every metre depth increase):
1,400,000/1,000 = 1,400 metres into the ocean, which is 1.4 km.

Question 6: B
Calculate that the probe can drop 300,000 Pa/1,000 Pa per metre = 300 metres into the ocean before breaking.
Now rearrange the equation in the question, to make t the subject, as follows:
$2d = \sqrt{(t^3)}$
$(2d)^2 = t^3$
$t = \sqrt[3]{(2d)^2}$
Then substitute the depth into this supplied equation:
$t = \sqrt[3]{(2d)^2} = \sqrt[3]{(2 \times 300)^2} = 71$ seconds.

SET 3

Question 7: A
Calculate the amount of drug taken for each disease:
Black Trump Virus = 4 mg x 80 kg x 3 times a day x 28 days = 26.88 g
Swamp Fever = 3x80x1x7 = 1.68 g
Yellow Tick = 1x80x2x84 = 13.44 g
Red Rage = 5x80x2x21 = 16.80 g
At a quick glance, the swamp fever dosage is much lower than all the others – you can discount this and use that to save a little time if you need to.

Question 8: A
First calculate that Carol took 20.16 grams of the drug during the two courses for Yellow Tick, using the same method as for John, but using Carol's weight of 60kg. Therefore 20.16 grams (the amount left over) corresponds to the dosage for the unknown disease:
4x60x3x28 = 20.16 g, therefore the unknown disease was Black Trump Virus.

Question 9: B
The first time he takes 3 x 80 x 1 x 7 = 1.68 grams, and the second time he takes 4 x 110 x 3 x 28 = 36.96 grams. Thus the ratio is 1.68 : 36.96 = 1:22.

Question 10: A
By calculating the dose required in each of the cases, we see that the only one that is above 15.5 grams over 4 weeks is the dosage for Red Rage:
5 x 75 x 2 x 21 = 15.75 g – therefore Danny must be suffering from Red Rage.

Question 11: D
Heavier people need a higher dose. To find the maximum weight, we use the equation: 5 x weight x 2 x 21 = 10 g, where "weight" represents the maximum weight requiring a dosage of less than 10 g.

So the maximum weight to not need a dosage exceeding 10 g is = 10,000 mg/(21 days x 2 daily doses x 5 mg/kg) = 47.62 kg.

SET 4

Question 12: B
To solve this, divide the flour content by the overall mass. A quick inspection might show you that this is likely to be Madeira, which is confirmed by the calculation (250/825 = 0.3). Thus, 30% of the Madeira's total weight is flour, which is a higher percentage than for any other cake

Question 13: A
In this question, there must be one cake where: (2,600/mass of cake) = (625/mass of flour in cake). Thus, there is a number that both the mass of the cake and the mass of flour can be multiplied by, in order to get these numbers respectively.
We can see that if we multiply the mass of the sponge cake by 5, we get 2,600 g. Equally, if we multiply the mass of flour in the sponge cake recipe (125g) by 5, then we get 625 g. Thus, Sponge cake is the answer. No other cake recipe can be multiplied by a given number to get an overall weight of 2,600 g and 625 g of flour.

Question 14: B
We use 1.50+1.25+1.10+1 times the ingredients for one cake, so the wedding cake will use 4.85 times as much of the ingredients listed for one cakes. We can use this to find which of the possible answers can be the amount of sugar in the cake, i.e. the sugar called for in one recipe multiplied by 4.85.
The quickest way to do this is to divide each possible answer by 4.85, and see if the result matches the weight of sugar in any of the cakes. We see that 970 g/4.85 = 200 g, which is the amount of sugar in the chocolate cake. None of the other amounts are possible. Thus, B is the answer.

Question 15: B

A kilogram of flour costs 55 x 2/3 pence and we are using 0.25 kg, so 9.167 p worth of flour goes into a Madeira cake. For sugar, we have 0.175 kg x 70 p per kg = 12.25 p worth of sugar going into the cake.

The ratio is thus 9.167:12.25 = approx 0.75:1 = 3:4

Question 16: D

As before, the flour costs: 55 p per 1.5 kg x 2/3 x 0.2 kg = 7.3 pence.

The milk costs: 44 p per kg x 150 g/1000 g per kg = 6.6 pence.

Thus the ratio is 7.3:6.6 = 1:0.9 = 10:9.

SET 5

Question 17: B

In total, 108 people out of the 200 tested have the disease; this is 108/200 = 0.54. Thus, the answer is 54%.

Question 18: D

As the infection rate is different for men and women, the infection rates must be calculated separately and combined:

(231,768 x 0.53 women x 0.63) + (231,768 x 0.47 men x 0.45) = 126,406 to the nearest whole person.

Question 19: B

There are 45 men and 63 women in the test group who have the Kryptos virus. Thus 15 of the men and 45 of the women have visited Atlantis. As we now know that 60 people have visited Atlantis, we can see that 108 – 60 = 48 have not visited. Now we simply calculate 48 as a percentage of 108. 48/108 = 0.44. Thus, 44% of people testing positive for the Virus in Test A have *not* visited Atlantis.

Question 20: B

We can see that 20/45 men testing positive in Test A have also tested positive in Test B, so we assume that the rest were false positives stemming from the inaccuracies of Test A. We are told to assume the same proportion of false positives in the women tested, so we simply apply this fraction to the number of women testing positive in Test A. Thus, we simply calculate 63 x (20/45) = 28. Thus, we expect that 28 women actually have Kryptos Virus.

Question 21: B

In total 108 people tested positive under test A, and 49 of these tested positive under test B (using the data given in the last question). Therefore the percentage of people positive in Test A also testing positive in Test B is 49 out of 108, which is 45.4%.

SET 6

Question 22: B

The cost of the plan is 190 + 600 + 140 = £930 per day

Question 23: C

Firstly we need to find the two options that save the most money, aside from the one already stated. The two best options are to send material from Plant A to Store 1, and material from plant B to store 2. We can see from the table that these 2 options will be £30 per day cheaper than sending from Plant A to store 2, and plant B to store 1 (as with the current business plan).

The new total cost is 100 + 180 + 450 = £730. Thus the saving is (930 – 730) = £200. £200 is 22% of £930, so the percentage saving is 22% (to the nearest whole number).

SET 7

Question 24: B
The shop sold 512 books outside of the visit event (sum of sales in the table), and 106 at the event.
Thus the percentage at the event was $106/(512+106) = 17\%$.

Question 25: B
Firstly calculate the different revenues:
Non-fiction revenue = $(12+30)$ x£10 = £420
Fiction revenue = $(50+45+23+90+103+159)$ x£6 = £2,820
Then calculate the non-fiction percentage: $420/(2,820+420) = 13\%$.

Question 26: B
The weekly revenue is seven times the daily revenue.
Daily revenue from Fiction: $(50+45+23+90+103+159)$ x£6 = £2,820.
Daily revenue from Non-fiction 2x$(12+30)$ x£10 = £840.
Therefore weekly revenue is 7x(£28.20+£840) = £25,620.

Question 27: D
The shop's revenue is now £6 x $(100+90+23+90+103+159)$ + £10 x $(12+30)$ per day. This income equates to £3,810 per day and £26,670 per week.

Therefore the percentage difference is $26,670/25,620 = 1.04$, giving a 4% increase on the previous week.

SET 8

Question 28: B
2 nights in Venice in a 3 star hotel, 1 room (the children are exempt) = 2x1x3 = 6 euros
2 nights in Rome in a 3 star hotel, 3 people paying (the child aged 9 is exempt) = 2x3x5 = 30 Euros
2 nights in Padua in a 3 star hotel, 2 people paying (both children exempt) = 2x2x2 = 8 euros
2 nights in Siena in the high season, 3 paying (child aged 9 exempt) = 2x3x2 = 12 euros
So the entire cost is the sum of these costs = 6+30+8+12 = 56 euros.

Question 29: B
In Rome he pays 6 euros x 7 nights = 42 Euros. In Padua he pays 3 euros x 8 nights = 24 Euros. 42:24 = 7:4.

Question 30: B
A 3 star hotel in Venice for 2 days costs Alice 6 Euros. 3 days in a 4 star hotel in Padua costs 9 Euros. This is 50% more in Padua than in Venice.

Question 31: C
The maximum cost of tax in a 4 star hotel in Rome is 6 euros x 10 nights = 60 EUR. Up until this point it will always be cheaper in Padua. In Padua the cost for a 4 star hotel is 3 euros a night, therefore after 20 days the cost of the Padua hotel is equal to the cost of a stay of equivalent duration in Rome.

SET 9

Question 32: B
First find the wall length: 15m x 0.6 = 9m. Therefore, the area of the room is 9m x 9m = 81 m^2

Question 33: B
The area of the house is $100 + 4 + (10 \times 20) + (3 \times 10) + (15 \times 15) + 81 = 640$ m^2.
The area of the master bedroom is 225 m^2, so the percentage area is $225/640 = 35\%$.

Question 34: D

The extra wall that is needed is the amount to extend both walls plus the new end wall.

The 20m wall is extended to 25m, and it is of height 3m, thus $5 \times 3 = 15$ m^2 is required.

Then the same amount of wall is needed for the other side, thus $2 \times 15 = 30$ m^2.

Then the back wall must be rebuilt, size = $10m \times 3m = 30$ m^2.

Sum these areas to a total of 60 m^2.

Question 35: B

Builder 1 costs: $(15 \times 300) + 200 = £4,700$

Builder 2 costs: $(16 \times 300) = £4,800$.

Therefore the requested ratio is 47:48, which is 1:1.02, to 3 s.f.

SET 10

Question 36: D

200 SMSs are free, so he pays for 7. £5 monthly fee + 7 SMSs x £0.20 + 15 minutes x £0.20 = £9.40

Question 37: C

Plan A SMSs cost 10p each. Plan B minimum cost is £5 monthly fee, but then provides 200 free SMSs. 51 SMSs at £0.10 each cost £5.10, which is more than £5. Thus, sending 51 SMSs would be cheaper on Plan B (sending 50 would be the same price on each plan).

Question 38: D

To answer this question, work out the cost of each plan.

5 minutes x 30 days = 150 minutes. 150 minutes cost: 150 x £0.10 = £15 on Plan A, 150 x £0.20 = £30 on

Plan B, (150-100 free minutes) x £0.20 = £10 on Plan C and 150 x £0.00 = £0 on Plan D.

Including the monthly fee, the cost to call becomes A: £15, B: £35, C: £20 and D: £15. All SMSs cost on Plan A but not D, so D is cheapest.

Note that we did not have to calculate the cost of texts in order to get the answer. If you like, you could work out the cost of the SMSs on each plan too, but if you notice that after calculating the cost of calls and monthly fees Plan A and Plan D come out at the same price, it is apparent that Plan D will be cheaper as SMSs are free, and this will save time.

Question 39: E

They each send 223 SMSs and make no calls. Option E is cheaper for Evan than Chris: For Plan B (223 SMSs – 200 free SMSs) x £0.20 = £4.60, so with the £5 monthly fee the total cost is £9.60. For Plan C all 223 SMSs are within the free limit, but the monthly fee is £10, so Chris pays more.

In option A, Evan pays £10.80, whilst Chris pays just the £10 monthly fee.

In option B, Evan pays £10.20, whilst Chris still just pays the monthly fee of £10.

In option C, Evan pays £36.80, whilst Chris still pays only the monthly fee of £10 (this is the biggest saving Chris makes relative to Evan out of all the options).

In option D, Evan pays between £27.40 and £29.80, depending on which month (and whether there are 28, 30 or 31 days in it), whilst Chris pays between £12.40 and £14.80, again dependent on which month.

Question 40: C

100 minutes are free on Plan C. Plan D has unlimited minutes and costs £5 more than C per month. At 20p a minute, Rachel can call for 25 minutes for £5, so can exceed her free minutes by 25 %, at which point Plan D is the same price. If she exceeds the free minutes allowance by any more than this, then plan D would be cheaper.

SET 11

Question 41: C

4 tablespoons is 4 x 15 ml = 60 ml. 250 ml is 1 cup, so 4 tablespoons is 60 ml/250 ml = 0.24 cups. 2 cups + 1 cup + ½ cup + 0.24 cups = 3.74 cups.

Question 42: A

First calculate the weight of butter called for by the recipe:
4 tablespoons = 60 ml = 0.06 litres = 0.06 dm^3 (the question states that $1dm^3$ = 1 litre).
Weight of butter: 950 grams/dm^3 x 0.06 dm^3 = 57 grams.

Next, calculate the weight of milk called for by the recipe:
½ cup = 1.25 dl = 0.125 litres = 0.125 dm^3.
Weight of milk: 1050 grams/dm^3 x 0.125 dm^3 = 131.25 grams.

Question 43: E

Any whole number multiple of flour, sugar or milk can be measured with a ½ cup.

4 tablespoons is 60 ml, ½ cup is 125 ml. The least common multiple of 60 and 125 is 1500, representing the smallest possible amount of butter that can be measured with half-cup measures.

1500 grams of butter makes 25 batches of muffins, which would require 25 x 2 cups = 50 cups of flour.
Thus, the weight ratio of Milk:Butter is 131.25/57:1 = 2.3:1.

Question 44: A

The recipe will have:
2 cups of milk = 500 ml = 0.5 dm^3
1 cup of sugar = 250 ml = 0.25 dm^3
½ cup of flour = 125 ml = 0.125 dm^3
4 tablespoons of butter = 60 ml = 0.06 dm^3.
Therefore the volume of the batter is 0.5 dm^3 + 0.25 dm^3 + 0.125 dm^3 + 0.06 dm^3 = 0.935 dm^3.

Now, to calculate the density of the batter, multiply the density of each ingredient by the proportion of the batter it makes up, and then add up these figures, as follows:
(1050 grams/dm^3 milk x 0.5/0.935) + (850 grams/dm^3 sugar x 0.25/0.935) + (600 grams/dm^3 flour x 0.125/0.935) + (950 grams/dm^3 butter x 0.06/0.935) = 561.5 grams/dm^3 + 227.3 grams/dm^3 + 80.2 grams/dm^3 + 61.0 grams/dm^3 = 930 grams/dm^3

Question 45: E

10 muffins x 100 grams = 1,000 grams batter required.

The recipe calls for:
2 cups flour = 5 dl = 0.5 dm^3 flour. Weight of the flour is 0.5 dm^3 x 600 grams/dm^3 = 300 grams.
1 cup sugar = 2.5 dl = 0.25 dm^3 sugar. Weight of the sugar is 0.25 dm^3 x 850 grams/dm^3 = 212.5 grams.
From question 42 we remember that the weight of the milk as called for by the recipe is 131.25 grams and weight of the butter is 57 grams.

Thus the overall weight of the batter is 300 + 212.5 + 131.25 + 57 = 700.75 grams.
700.75/1,000 grams batter = 57/B grams butter, where B is the amount of butter required in 1,000g batter. B = 81.3 grams butter.

Question 46: C

700.75/1,000 grams batter = 300/F grams flour, where F is the amount of flour required to make 1,000 grams of muffin. F = 428.11 grams. 428.11 grams/1000 grams = 0.428 = 43 % to the nearest whole number.

SET 12

Question 47: A

The trend shows a 10 km^2 decrease in thickness for every 100 °C decrease in temperature. 10 km^2 thickness is 1,300 °C, so extrapolating the trend gives 0 km^2 thickness at 1,200 °C.

Question 48: D

Spreading rate is not affected by temperature, it is an independent variable. This question is designed to test your attention to detail and reinforce the importance of reading questions properly. Ensure you constantly pay attention to what the question is asking.

Question 49: D

Crustal volume per year = 20 km^2 crustal thickness x 20 mm/year spreading rate x 1 year = 20,000,000,000,000 mm^2 x 20 mm = 400,000,000,000,000 mm^3 = 400,000 m^3

Question 50: C

In answering this question, it is not necessary to use the same units for crustal thickness and spreading rate, as long as we use the same units for the crustal thickness *from both locations*, and likewise for the spreading rate.

Location A = 10 x 100 = 1,000. Location C = 30 x 150 = 4,500. A:C = 1:4,500/1,000 = 1:4.5

Question 51: A

Crustal volume per time = crustal thickness x spreading rate.
Crustal thickness at E is 10 km^2 and at F is 25 km^2.
Crustal volume per time is equal, so 10 km^2 x spreading rate E = 25 km^2 x spreading rate F. Therefore the spreading rate at E = 2.5 the spreading rate at F. Thus, it is 250 % faster.

Question 52: E

Temperature at D is 1,600 °C, decreased by 10 % it would be 1440 °C.
From the trend in temperature and crustal thickness this corresponds to a crustal thickness of 24 km^2.
Crustal volume per 3 years = 24 km^2 crustal thickness x 50 mm/year x 3 years = 24,000,000,000,000 mm^2 x 150 mm = 3,600,000,000,000,000 mm^3

SET 13

Question 53: C

A: 9/25, B: 6/23, C: 7/22, D: 8/24. (9+6+7+8)/(25+23+22+24) = 30/94 = 32 % to the nearest whole number.

Question 54: E

In Group A, drug-takers visual accuracy is 36%/27% = 33.33% improved.
In Group C, drug-takers visual accuracy is 31%/29% = 6.90% improved.
A:C = 33.33%/6.90%:1 = 4.83:1.

Question 55: C

10 women and 15 men have 45 % and unknown (P %) accuracy, respectively.
All 25 have an average of 36 % accuracy.
(10/25 x 45) + (15/25 x P) = 36.
P=30.

Question 56: A

Diabetics with vision problems correspond to Group A. In Group A, 15 of 25 volunteers reported better vision after taking the drug, but 9 of 25 volunteers taking a placebo also reported vision improvement. This suggests only 6 of 25 had reported improvements in their vision thanks to the effects of the drug.
6/25 x 100,000 people = 24,000 people.

Question 57: D

The placebo group showed no change, so only the volunteers being affected by the drug compound will see greater improvements to their vision.

15 – 9 = 6 volunteers affected by drug originally.
200% of the dose gave (18-9)/6 = 1.5 = 150 % the number of people with drug-related improvements.
300% of the dose will thus give 300 % / 200 % x 150 % = 225 % the number of people.
6 volunteers x 225% = 13.5 volunteers. (9+13.5)/25 volunteers = 22.5/25 volunteers = 90.0 % of volunteers.

Question 58: D
D is supported, because in all groups taking the placebo, there was an increase in accuracy with reading letters, suggesting better vision, and in many cases this was equivalent to the increase in those taking the drug.

The data suggest that A) and B) are incorrect. Only one group showed a higher increase in accuracy amongst the placebo group, in all other groups the people taking the drug had a larger % increase in accuracy. In healthy volunteers, there was as much of an increase in accuracy amongst those taking a placebo, suggesting the drug does not have as much of an effect in healthy volunteers. Options C) and E) are relatively meaningless statements which are not supported by the data.

SET 14

Question 59: C
5 calories x 200 pounds x 1 hour = 1000 calories running.
Cycling burns 50 calories + (5 calories x -5) for each mile = 25 calories per mile. Cycling 5 miles gives a total burn 125 calories, which is less than the amount burned running.

Thus, the maximum calorie burn comes from running, which will burn 1000 calories.

Question 60: D
Losing 10 pounds requires a 35 000 calorie deficit.
30 min run: 200 pounds x 0.5 hours x 5 calories = 500 calories.
20 mile cycle at 20 mph: 50 calories + (5 calories/mph x 10 mph) = 100 calories/mile for 20 miles = 2,000 calories. Daily burn is 2500 calories. 35,000/2,000 calories per day = 14 days taken to lose 10 pounds.

Question 61: C
Both their weight loss goal and calories burned running are linearly proportional to weight. Thus, they need to run for the same amount of time in order to achieve their goals.

Question 62: A
10% of 140 pounds is 14 pounds of weight, thus this is how much weight she wishes to lose.
At 3,500 calories/pound she needs a 14 pounds x 3500 calories/pound = 49,000 calorie deficit.

Her BMR is 1500 calories and she eats 400+500+250+200=1,350 calories per day, so her daily calorie deficit is 1,500 calories-1,350 calories = 150 calories.

It will take her 49 000 calories/150 calories per day = 326.67 days to reach her goal (327 to the nearest day).

Question 63: E
10 miles at 10 mph: 50 calories + (5 calories x 0 mph) = 50 calories/mile for 10 miles = 500 calories.

She requires a 49,000 calorie deficit, as worked out in the previous question. Previously she had a daily deficit of 150 calories, now she has a daily deficit of 650 calories. Thus, she now has a calorie deficit which is 4.33 times the previous deficit, so she will reach her goal 4.33 times faster.

Question 64: E
1 chocolate is the least she can eat, which means she eats 3 pieces of chicken and 6 bowls of cereal.
1x350 calories + 3x250 calories + 6x400 calories = 3,500 calories.

The lowest calorie arrangement of the 3 other foods is 1 lasagna, 3 vegetables and 6 apples.
1x700 calories + 3x200 calories + 6x100 calories = 1,900 calories.

She has a 3,500 calories per day – 1500 calorie BMR = 2,000 calorie surplus before and 400 calorie surplus after. 2,000 calories:400 calories = 5:1.

SET 15

Question 65: D
Price during day with single tickets is 4 coupons x £1 = £4.
Price at night with 10% off coupons is 3 coupons x £0.90 = £2.70.
Thus the saving is £1.30
Hence, the % saving is £1.30/£4.00 = 0.325 = 32.5 %.

Question 66: B
Price without a wristband is £1 per coupon x 120 % = £1.20 per coupon.
The number of coupons required is: 10 coupons entrance + 40 coupons for roller coaster + 6 coupons for candy floss + 1 coupon for the games + 4 coupons for the fun house + 3 coupons for the swings = 64 coupons.
64 coupons x £1.20 = £76.80.
Price with a wristband: £70 and 6 coupons for candy floss + 1 coupon for the games x £1.20 = £8.40. Thus the total price with a wristband is £78.40.

Therefore, the ratio is 1:£76.80/£78.40, which is 1:0.98.

Question 67: C
Andy rode the swings 20 % more times than he played games. The smallest number of times he could have gone on the swings is 6 times (since he can't have gone on the swings a non-integer number of times) so he played 5 games and went on the rollercoaster 9 times.
At night rates, rollercoaster 9 times x 3 coupons = 27 coupons and swings 6 times x 2 coupons = 12 coupons.
39 coupons at £1 each are worth £39, which is less than the £70 wristband.

If he had played 10 games, ridden the swings 12 times and rollercoaster 18 times, the wristband would have been more cost effective. This is the second cheapest possibility whilst still going on the swings an integer number of times (since 10 is the next multiple of 5). Thus, we know Andy must have played 5 games, been on the swings 6 times and the rollercoaster 9 times.

39 ride coupons + 5 game coupons + 5 entrance coupons = 49 coupons = £49.

Question 68: D
They got 69 coupons each for £69 pounds. They each must have spent 10 coupons at the entrance.
Together the roller coaster and fun house cost 6 coupons, so Anna took 59/6 = 9 rides on each and had 5 coupons left over, since the question states she went on each ride for *half* her rides, we assume she did not use any of the remaining coupons, since she could only have gone on one more ride, and thus it would not have been exactly half.

The fun house and swings cost 5 coupons together, so James took 59/5 = 11 rides on each and had 4 coupons left which he spent on 1 last ride on the swings.

Anna took 18 rides and James 23 rides. Anna: James = 1:23/18 = 1:1.28.

Question 69: E

First we calculate how much candyfloss Erik buys each weekend, using the 100% increase for the first 4 weeks, and the 50% increase thereafter as detailed in the question.

Week 1: 1, 2: 2, 3: 4, 4: 8, 5: 16, 6: 24, 7: 36, 8: 54, 9: 81, 10: 0. Thus, he bought a total of 226 lots of candy floss.

226 candy floss x 2 coupons x £1 pound each = £452.

Cost with season pass = £1000.
Cost without pass = £700 + £452 = £1,152.

SET 16

Question 70: B

2 x (£14.00 + £2.00 + 2x£2.00) + 1 x (£8.00 + 3x£1.00) + 3 x (£10.00) = £40.00 + £11.00 + £30.00 = £81.00.
50% off orders over £50 means he pays £81.00/2 = £40.50.
Ratio of cost Without:With is £1152/£1000:1 = 1.15:1.

Question 71: C

£35 is 50% of £70.
£70 buys £70/£14 = 5 large cheese pan pizzas.
Large pizzas have 10 slices: 10 slices x 5 pizzas = 50 slices.

Question 72: E

£60 is 50% off £120.
£120 is evenly divisible by all prices of plain cheese pizzas except the large pan. Dividing £120 by the price of the large pan gives 8.571. Since Joey cannot have bought a non-integer number of pizzas, he cannot have bought this type of pizza which costs £14.00.

Question 73: D

30 slices is 3 large pizzas or 5 small pizzas.
3 large pan pizzas with 2 toppings and stuffed crust cost: 3 x (£14.00 + (2x£2.00) + £2.00) = £60.00.
5 small pan pizzas with 2 toppings and stuffed crust cost: 5 x (£10.00 + 2x£0.50 + £1.00) = £60.00.
Large:Small ratio is 1:£60.00/£60.00 = 1:1.00

Question 74: A

The lowest price order over £31 is 2 small pizzas and 1 medium pizza costing £32 total. 30% off £32 = £22.40. They get 20 slices with this order, but we know that they got 25% more than they could eat. Thus, if we treat 20 slices as 125%, then 100% is the amount that they ate. 20/1.25 = 16, so this is the amount that they could eat.

SET 17

Question 75: E

1 hour at Job A = £10 starter - £5 travel + £10 per hour = £15.
1 hour at Job B = £5 starter - £0 travel + £15 per hour = £20.
1 hour at Job C = £5 starter - £10 travel + £20 per hour = £15.
A:B:C = £15:£20:£15 = 1:1.33:1.

Question 76: C

In her first 50 hours she worked 25 2-hour jobs: 25 x (£10 - £5 + 2x£10) = 25 x £25 = £625.
Her hourly wage was then increased to £10 + £5 = £15.

She then worked 4 1-hour jobs: 4 x (£10 - £5 + 1x£15) = 4 x £20 = £80.
Then she worked 1 4-hour job: 1 x (£10 - £5 + 4x£15) = 1 x £65 = £65.

Her total earnings were £625 + £80 + £65 = £770.00

Question 77: E
For Job A, for the first 50 hours she would earn £10 an hour. Thus, for the first 25 2-hour jobs she would earn 25 x (£10 - £5 +2x£10) = £625. After 50 hours, her hourly wage would increase to £15 an hour. Thus, for the second lot of 25 2-hour jobs she would earn 25 x (£10 - £5 + 2x£15) = £875. Thus, in total she would earn £1,750.
For Job B, for the 50 2-hour jobs, she would earn 50 x (£5 + 2x£15) = £1,750. She does not get the pay increase because she does not work any hours after the pay rise at 100 hours.
For Job C, for the 50 2-hour jobs, she would earn 50 x (£5 - £10 + 2x£20) = £1,750.
Thus, B and C are the same.

Question 78: D
Average job length for Job C is 4 hours.
1 4-hour job earns: £5 starter - £10 travel + 4x£20 per hour = £75 per job.
Maximum earnings £1250/£75 = 17 jobs.
17 jobs x 4 hours = 68 hours.

Question 79: D
Hourly wage has risen twice by £5 to £30 per hour (since it rises £5 for each 100 hours worked).
100 hours/4 hour jobs = 25 jobs.
For a 4 hour job she earns: £5 starter - £10 travel + 4x£30 per hour = £115 per job.
25 jobs x £115 per job = £2,875.

10% of £2,875 = £287.50. This is the amount she pays in income tax.

Question 80: D
Working 50 hours for Job B earns 50 times the hourly rate plus 50 times the fixed starter wage, as jobs average 1 hour in duration. Therefore the income = (50x£15) + (50x£5) = £1,000.00.
The extra 50 hours can be worked at either Job A, B or C (note that Jobs A and B have travel expenses).
The same calculation for Job A gives (10x50) + (10 x 25) – (5x25) = £625.00.
The same calculation for Job C gives (50x20) + (5x12.5) – (10x12.5) = £937.50.

Therefore she earns most by working 100 hours for Job B.

SET 18

Question 81: A
Monday has the 2nd highest number of passengers, so this will be the 2nd highest grossing day. On Monday the revenue is £5 x 2346 underground passengers = £11,730.00.
On Wednesday the revenue is 3,103 x 0.85 x £5 = £1,3187.75. £13,187.75 - £11,730.00 = £1,457.75

Question 82: C
Total number of weekday underground passengers = 2346 +1798 +3103 + 2118 + 1397 = 10,762
Total number of weekday car passengers = total passengers – number of drivers =
Monday: 1873-1517=356
Tusesday: 2421-1632=789
Wednesday: 1116-987=129
Thursday: 2101-1465=636
Friday: 2822-2024 =798
356+789+129+636+798 = 2708

The average ratio is 10,762 : 2,708 = 1:3.97

Question 83: E
Tuesday: 2,421/1,632= 1.48
Weekend: 1,339/478= 2.80.
Ratio of Tuesday to Saturday = 1.48/2.8 = 1:1.89

Question 84: B
This is straightforward: 2,346/576 = 4.07:1 = 1:0.25

Question 85: D
There are 4,219 commuters each day (as seen by adding the total car passengers and underground passengers on any given day). For every 1 car there are 1.7 passengers and 2 underground riders. Thus, by dividing 1.7 by 3.7 and multiplying this by the total number of passengers (4219), we can calculate how many people are in cars:
1.7/3.7 x 4,219 = 1,938 people drive (to the nearest whole number, obviously there cannot be a non-integer number of people driving).
Now, by dividing the number of people driving (1938) by the number of passengers, we can calculate the number of cars: 1938/1.7 = 1,140 cars.
1,140 cars x £4 = £4,560. £4,560 + 1,938x£1 = £6,498. 80% of £6,498 is £5,198.40.

SET 19

Question 86: E
First calculate the Superior room cost at night: 3hrs x £26.00/hr = £78.00
Now add 10%: £78 x 1.1 = £85.80

Question 87: D
Total cost minus deposit = 12hrs x £30/hr = £360
Deposit = £460 - £360 = £100

Question 88: B
Total deposit = £10 + £25 = £35
6hrs in Standard room = 6hrs x £18/hr = £108
6hrs in Basic Room = £221 – (£108 + £35) = £78
Basic room hourly rate (2-6hrs) = £78 ÷ 6hrs = £13/hr

Question 89: B
1.5hrs in Superior room (night session) = 1.5hrs x £30/hr = £45
Decreased by 5% = £45 x 0.95 = £42.75
Total cost = £50 deposit + £42.75 = £92.75

Question 90: B
Basic room all day cost = 18hrs x £8/hr = £144
All week = 7 days x £144 = £1,008
Three weeks = £1,008 x 3 = £3,024
Deducting the VAT = £3,024/1.25 = £2,419.20

SET 20

Question 91: C
Total number of people aged under 22 = sum of first two columns = 62
Number aged <22 who spotted >10 differences = 11 + 8 + 3 + 2 = 24
Percentage = 24/62 = 38.7%

Question 92: D
Valid results for 5-10 spots for ages 16-22 = 0.25 x 12 = 3
Total number of valid results for 16-22 = 10 + 3 + 8 + 2 = 23
Percentage of over 15 spots for 16-22 = 2 ÷ 23 = 8.7%

Question 93: C
Total who spotted over 10 = sum of bottom two rows = 52
25% of 52 = 13

Question 94: E
Total 48+ who spotted <5 = 15 + 19 = 34
Total aged 48+ = sum of final two columns = 66
Percentage of 48+ who spotted <5 = 34 ÷ 66 = 52% (2 s.f.)
52% of 10,000 = 0.52 x 10,000 = 5,200

Question 95: E
50% increase in 16-34s who spot 11-15 = 1.5 x (8 + 6) = 21
New total who spot 11-15 = 11 + 21 + 2 + 8 + 9 = 51
Ratio = 21:51

SET 21

Question 96: B
Number that play Football = 22% of 1300 = 286
Number that play Hockey = 8 % of 1300 = 104
Thus, the difference = 286-104 = 182 Boys + Girls; Therefore, 182/2 = 91 Boys

Question 97: D
22% of 350 = 77 students
77 ÷ 11 people per team = 7 teams

Question 98: C
Number of basketball boys = 0.05 x 1,300 = 65
80% of 65 = 52
Number of netball girls = 0.08 x1, 300 = 104
Total number of netball-players = 104 + 52 = 156
Male proportion = 52 ÷ 156 = 33% (2 s.f.)

Question 99: D
Number of *Other* students = 0.12 x 1,300 = 156
Other ball sports = 0.25 x 156 = 39
Total non-ball sports (swimming & athletics) = (0.06 + 0.03) x 1300 = 117
Total "other" non ball sports = 156 – 39 = 117
Total ball sports = 1,300 – 117 – 117 = 1,066

Question 100: A
Girls who play hockey = 0.07 x 1,300 = 91
Boys who play cricket = 0.1 x 1,300 = 130
Difference = 39. *Note the tennis info makes no difference (50:50 split)*

SET 22

Question 101: A
Total apples processed in 1998 = 1,100,547 + 2,983,411 = 4,083,958
Total apples processed in 2003 = 1,931,784 + 2,439,012 = 4,370,796
Ratio = 4,370,796 ÷ 4,083,958 = 1.07 (i.e. 7% increase)

Question 102: E
Number of No Goods in worst year = 571,221
Total number of No Goods = sum of bottom row = 2,823,732
Percentage = 571,221 ÷ 2,823,732 = 20.2% (3 s.f.)

Question 103: A
2004 total No Goods = 3 x 571,221 = 1,713,663
70% of edible apples are processed, as are all passable apples.
Difference = (0.7 x 1,931,784 + 2,439,012) − 1,713,663 = 2,077,598

Question 104: C
Total number of edibles 1998-2003 = sum of top row = 9,201,790
20% increase = 1.2 x 9,201,790 = 11,042,148
30% of these are sold as they come = 0.3 x 11,042,148 = 3,312,644

Question 105: B
Apples processed for cider = (1,931,784 x 0.7) + 2,439,012 = 3,791,260.8
Litres of cider = 3,791,260.80 ÷ 20 = 189,600 litres (4 s.f.)

SET 23

Question 106: C
Decreased average speed = 0.92 x 5 mph = 4.6 mph
Miles covered = 4.6 x (40 ÷ 60) = 3.07 miles
Km covered = 1.6 x 3.07 miles = 4.9 km

Question 107: C
Distance per session = 26 miles ÷ 4 = 6.5 miles
Time per session = 6.5 miles ÷ 5 mph = 1.3 hrs = 1hr 18mins

Question 108: B
New wet average speed = 0.92 x 4.6 mph = 4.232 mph
12 km in miles = 12 km ÷ 1.6 = 7.5 miles
Time taken = 7.5 miles ÷ 4.232 mph = 1.77 hrs = 1hr 46mins

Question 109: B
Distance of second jog = 4 km x 1.5 = 6 km
Distance of third jog = 6 km x 1.5 = 9 km
Distance of last jog = 9 km x 1.5 = 13.5 km
13.5 km in miles = 13.5 km ÷ 1.6 = 8.4375 miles
Time = 8.4375 miles ÷ 5 mph = 1.69 hrs = 1hr 41mins

Question 110: C
3hrs 42mins = 3.7 hrs
Average speed = 26 miles ÷ 3.7 hrs = 7.027 mph
7.027 mph ÷ 5 mph = 1.405 (i.e. 41% increase)

SET 24

Question 111: C
Standard price = (£325 + £100) ÷ 2 = £212.50
For two people for one week = £212.50 x 2 = £425
Deducting 20% due to discount = 0.8 x £425 = £340

Question 112: C
Four weeks rent per person = 4 x £480 = £1,920
Total group rent = 12 x £1,920 = £23,040
Booking fee per person = 0.1 x £480 = £48
Total booking fee = 12 x £48 = £576
Total cost = £23,040 + £576 = £23,616

Question 113: C
One person for 10 days = £80 x (20÷7) = £228.57
Two people for 10 days = 2 x £228.57 = £457.14
Booking fee = £492.89 - £457.14 = £35.75

Question 114: A
Total rent = £220 x 2 weeks x 4 people = £1,760
Total discount = £1,760 x 0.2 = £352
Total cost = £1,760 - £352 = £1,408

Question 115: A
Booking fee = 0.1 x £19,500 = £1,950
Total rental charge = £19,500 - £1,950 = £17,550
Number of people = £17,550 ÷ £325 = 54
Number in each palazzo = 54 ÷ 3 = 18 people.

SET 25

Question 116: B
The price of the food is £3.95 + £2.95 + £3.95 = £10.85. Orders between £10 and £15 are charged £1.50 for delivery. Including delivery, the cost is £12.35. We then add 20%, so multiply this by 1.2. Hence the cost of the takeaway is £14.82.

Question 117: C
The total price is (3 x £2.95) + (2 x £4.95) + £3.95 = £22.70. Delivery is free above £15. We then add 20%, hence the total cost is £27.24.
A promotional offer is introduced whereby customers receive a 10% discount for orders over £9 (not including delivery).

Question 118: A
The total price is (2 x £2.95) + £3.95 = £9.85
Add £3 for delivery under £10: 9.85 + 3 = 11.865
Add 20% for VAT: 11.865 x 1.2 = £15.42

Question 119: B
The total is £3.95 + (3 x £2.95) + £3.95 = £16.75
Delivery is free over £15. We then add 20% for VAT: 16.75 x 1.2 = £20.10

Question 120: D
The total price is (2 x £2.95) + (2 x £4.95) + £3.95 = £19.75 (price of only two noodles because of offer)
Delivery is free over £15. We then add 20% for VAT: 19.75 x 1.2 = £23.70

SET 26

Question 121: B
Divide the profits earned by MediCo by the total profits earned by all suppliers: (15,000 + 30,000 + 25,000 + 35,000) = £105,000. 30,000/105,000 = 28.6%.

Question 122: C
Divide the profits earned by MediCo and Lifecare combined by the total profits earned by all suppliers: (15,000 + 30,000 + 25,000 + 35,000) = £105000. (30,000 + 35,000)/105,000 = 61.9%.

Question 123: C
Increase the profit of PillPlus by 15% and use this value when calculating the new total profit earned by all suppliers: 25,000 x 1.15 = 28,750. 28,750/(15,000 + 30,000 + 28,750 + 35,000) = 26.4%.

Question 124: A
As all profits across suppliers decrease by an equal amount, this is irrelevant and cancels in the division. 0.9(35,000 + 15,000)/0.9(35,000 + 28,750 + 30,000 + 15,000) = 46%.

Question 125: B
Take account of the first fall in profits of 10% and then a further fall of 5%: (15,000 + 35,000) x 0.9 x 0.95 = £42,750.

SET 27

Question 126: A
The total value for the Stuntman is £12500 + £345 + £145 + £295 = £13,285.
Adding 20% VAT, this is £13,285 x 1.2 = £15,942.

Question 127: B
Remember to include tax in all calculations:
Saloon: (£21500 + £495 + £245 + £445) = £22,685. £22,685 x 1.2 = £27,222.
Pod: (£18000 + £445 + £395 + £495) = £19335. £19335 x 1.2 = £23,202.
£27,222 - £23,202 = £4020

Question 128: C
Racer: (£15,000 x 1.2) = £18,000.
Stuntman: (£12500 + £345 + £145 + £295) = £13,285. £13,285 x 1.2 = £15,942.
£18,000 - £15,942 = £2,058

Question 129: C
The Pod with no optional extras is £18000 x 1.2 = £21,600.
The Racer with all optional extras with a 10% discount is (£15000 + £395 + £195 + £395) x 0.9 x 1.2 = £17,263.80.
£21,600 - £17263.80 = £4,336.20

Question 130: D
Saloon with leather seats and easy-park technology = £21500 + £495 + £445 = £22440.
Add 20% = £22,440 x 1.2 = £26,928. Reduced by 20% = £26928 x 0.8 = £21,542.40.
Pod = £18,000 x 1.2 = £21600.
£21,600 - £21,542.40 = £57.60

SET 28

Question 131: C
Emissions increased by 1,000 tonnes from 1,000 to 2,000 tonnes over 5 years, therefore the rate of increase was 200 tonnes per year

Question 132: A
If there had been no crash, 2,010 emissions would have been 3,000 Tonnes, as calculated by applying an increase in emissions of 200 Tonnes/year from 2005 to 2010.
With the economic crash, 2010 emissions were 2500 Tonnes:
3,000 – 2,500 = 500 Tonnes less in 2010 due to the economic crash.

Question 133: C
Percentage increase = (new amount – old amount)/old amount x 100
= (3,000–2,000)/2,000 x 100 = 50%.
Note that you are asked for the percentage INCREASE. (New amount/old amount) x100 = percentage CHANGE.

Question 134: B
2015 – 2020 the amount would increase from 3,000 tonnes to 3,500 tonnes without any action. This equates to a rate of increase of 100 tonnes per year. With the new act, this is reduced by 50% to 50 tonnes per year, thus over 5 years: Overall saving = 50 x 5 = 250 Tonnes

Question 135: C
The emissions in 2020 = projected – rate reduction (50% of the difference between projected total and current total). Rate reduction = 50 tonnes per year. Projected rate = current emissions (3,000 tonnes) + projected increase without the act (5 x 100, which is 500).
= (3,000+500) – (5 x 50) = 3250 Tonnes

SET 29

Question 136: B
For between 10 and 100 units, single sided black and white printing costs £0.07 per page. 74 x 7p = £5.18.

Question 137: C
£100 will clearly buy more than 100 sheets as all prices are under £1, therefore use the 100+ price:
The price for double sided colour printing over 100 sheets is £0.25.
£100/0.25 = 400 sheets

Question 138: A
For over 100 sheets of double sided black and white, the cost per sheet is £0.10 each. Hence the total non-discounted price for 150 sheets is 150 x £0.10 = £15. If this is discounted by 10%, this is £15 x 0.9 = £13.50.

Question 139: B
Price for buying separately = 150 x 0.15 = £22.50
Price buying together = 150 x 0.10 x 0.9 (discount) = £13.50
Difference = £9
Number of sheets could buy with £9 = 9/0.15 = 60 sheets

Question 140: B
The price for 227 sheets of double sided black and white printing is £0.10 each, so the total is £22.70.
The price for 34 sheets of double sided colour printing is £0.30 each, so the total is £10.20.
The total for all the sheets is £32.90.
If this is discounted by 10%, this is £32.90 x 0.9 = £29.61.

SET 30

Question 141: D
We can calculate the number of people who reacted positively in each group and then add up the total:
75% of 300 is 225
65% of 300 is 195
70% of 300 is 210
55% of 300 is 165
225 + 195 + 210 + 165 = 795

Question 142: C
In group 2, 30% reacted negatively: 30% of 300 is 90.
In group 3, 15% reacted negatively: 15% of 300 is 45.
Therefore the difference is 45; 45 more people in group 2 reacted negatively than people in group 3.

Question 143: B
We can calculate the number of people who reacted negatively in each group and then add up the total:
20% of 300 is 60
30% of 300 is 90
15% of 300 is 45
25% of 300 is 75
60 + 90 + 45 + 75 = 270
270 as a percentage of the total, 1200, is 22.5%, which rounds to 23%.

Question 144: D
The overall success rate of the first four groups was = (75 + 65 + 70 + 55)/4 = 66.25%
Therefore increase in success rate = 82/66.25 = 23.77% increase.

Question 145: B
In the answer to question 143 we calculated that in the first 4 groups, 270 people reacted negatively. In group 5, 15% of 300 people, which is 45 people, reacted negatively. Hence the negative reactions total 315 people.

SET 31

Question 146: C
The total number of views of The Last Chase is 20,000 + 20,000 + 15,000 + 20,000 = 75,000.
The total number of views of The Final Frontier is 15,000 + 20,000 + 25,000 + 35,000 = 95,000.
Hence the difference is 95,000 – 75,000 = 20,000

Question 147: B
The difference in viewers is 20,000. If £2,500 is earned per 1,000 viewers, then the difference in advertising revenue will be £2,500 x 20 = £50,000.

Question 148: B
Growth rate is 5,000 per quarter between Q1 and Q3 of 2014. Therefore the total growth during 2015:
35,000 + (5,000 x 4) = 55,000

Question 149: B
Growth rate is 5,000 per quarter between Q1 and Q3 of 2014. Therefore the total growth during 2015:
Q1: 35,000 + 5,000 = 40,000
Q2: 40,000 + 5,000 = 45,000
Q3: 45,000 + 5,000 = 50,000
Q4: 50,000 + 5,000 = 55,000
Thus, the total number of views = 40,000 + 45,000 + 50,000 + 55,000 = 190,000.

Question 150: D
Without the show cancellation, there are 35,000 Final Frontier views and 20,000 Last Chase views. With the cancellation, half of the Last Chase views transfer over to Final Frontier, this equates to 10,000 views. Therefore the Final Frontier now has 45,000 views in this scenario.

SET 32

Question 151: C
55% + 25% = 80% of Scottish patients wait less than half an hour. 80% of 50,000 patients = 40,000, therefore 40,000 patients waited for less than half an hour for an appointment.

Question 152: B
We can work out how many patients had to wait more than half an hour for an appointment in each part of the UK then find the total:
In England, 10% of 100,000 patients = 10,000 patients had to wait longer than half an hour.
In Scotland, 20% of 50,000 patients = 10,000 patients had to wait longer than half an hour.
In Wales, 25% of 25,000 patients = 6,250 patients had to wait longer than half an hour.
In Northern Ireland, 15% of 25,000 patients = 3,750 patients had to wait longer than half an hour.
The total number of patients that had to wait longer than half an hour is 10,000 + 10,000 + 6,250 + 3,750 = 30,000.
30,000/200,000 = 15%

Question 153: C
In the previous question we found that 30,000 patients had to wait longer than half an hour, so 40% of 30,000 people complained = 12,000.

In England, 30% of 100,000 patients = 30,000 patients had to wait 11-30 minutes.
In Scotland, 25% of 50,000 patients = 12,500 patients had to wait 11-30 minutes.
In Wales, 25% of 25,000 patients = 6,250 patients had to wait had to wait 11-30 minutes.
In Northern Ireland, 25% of 25,000 patients = 6,250 patients had to wait 11-30 minutes.
The total number of patients that had to wait 11-30 minutes is 30,000 + 12,500 + 6,250 + 6,250 = 55,000.
If 20% of 55,000 patients complain, this is 11,000.

Hence the total number of complaints is 11,000 + 12,000 = 23,000.

Question 154: B
Use the total survey size to calculate the proportion: 23,000/200,000 = 11.5%.

Question 155: B
We worked out in question 153 that 30,000 people had to wait longer than 30 minutes and 55,000 people had to wait between 11 and 30 minutes. If the targets are met, 15,000 people will have to wait longer than 30 minutes and 41,250 will have to wait between 11 and 30 minutes. If 20% of these 41,250 people complain, this is 8,250 complaints. If 40% of the 15,000 people complain this is 6,000 complaints. Hence there are a total of 14,250 complaints. The previous number of complaints was 23,000, hence the percentage decrease is (23,000-14,250)/23,000 = 0.38 = 38%.

SET 33

Question 156: D
The decrease in the price of crude oil between January and March is $150-$100 = $50, as can be seen in the graph. This fall has occurred over a 2-month period, giving a decrease in price of $25 per month.

Question 157: D
This is a question of estimation. The average production across the year is at least 7 million barrels per day. Multiplying this by 365 gives around 2,550 million barrels per year. All other options require less than 7 million barrels daily production, and it is clear there are at least 7 million barrels per day. Therefore the answer is 2,700 million.

Alternatively, we can estimate using 30 days per month, and multiplying the amount of barrels produced per day in each month by 30 (this is more accurate but more time consuming). 6+7+7+7.5+7.5+7+7.5+8+8.5+8.5+8+9 = 91.5, multiplying by 30 gives just over 2,700 million barrels per year.

Question 158: C
Use both graphs. For July, multiply the oil price by the amount sold in the month, and then multiply by the number of days in the month.
July = 7.5 million barrels x $75 per barrel x 31 days = $17,400 million = $17.4 billion

Question 159: B
If the costs are 40%, the gross profit is 60%.
Oil sales in June 2014 totalled 7 million barrels x $100 per barrel x 30 days = £21,000 million = $21 billion.
Therefore gross profit was 0.6 x $21 billion = $12.6 billion.

Question 160: A
You are given the total sales value of $204 billion, so work with this. Work clearly in stages and this question is not hard.
The profit is 60% of this, which is $122.4 billion.
This is split 5:2 between the oil companies and the oil-producing nation. Thus, the profit for the oil companies is 5/7 of $122.4 billion is profit for the oil companies, which is $87.43 billion.
Corporation tax is then 30% of this profit, which is $26.23 billion.

SET 34

Question 161: B
Calculate the total people with no asthma, then take it away from the total number of people which is 250:
80% of the 50 people aged 0-5 have no asthma, which is 40.
75% of the 50 people aged 5-10 have no asthma, which is 37.5.
85% of the people aged 10-21 have no asthma, which is 42.5.
95% of the people aged 21-30 have no asthma, which is 47.5.
95% of the people aged 30+ have no asthma, which is 47.5.
Hence the total people who have no asthma is 215.
Hence the total people with asthma is 250 – 215 = 35.

Question 162: B
Of children aged 0-5, 15% have mild asthma which is 7.5. Hence 3.75 will develop respiratory problems.
Of children aged 0-5, 5% have severe asthma which is 2.5. Hence 2.25 will develop respiratory problems.
Of children aged 5-10, 20% have mild asthma which is 10. Hence 5 will develop respiratory problems.
Of children aged 5-10, 5% have severe asthma which is 2.5. Hence 2.25 will develop respiratory problems.
Hence 13.25 children per 100 will develop respiratory problems, so the answer is 13.25%.

Question 163: A
(15%+20%)/2 = 17.5% diagnosed. 17.5% x 0.35 = 6.125% Incorrect diagnoses.

Question 164: B
Taking into account that only 65% of children diagnosed with mild asthma were diagnosed correctly:
17.5% diagnosed (mild) ➜ 11.375% correctly diagnosed ➜ 5.69% complications
5% diagnosed (severe) ➜ 4.5% complications (all are correct diagnoses)
Therefore 5.69 + 4.5 = 10% overall respiratory complication rate.

Question 165: A
False diagnoses: (0.15x0.35x0.07x50,000,000) 0-5 year olds + (0.2x0.35x0.1x50,000,000) 5-10 year olds = 533,750
Cost per diagnosis: £50 per year = £250 over 5 years
Total money wasted: 533,750 x £250 = £133 million

SET 35

Question 166: D
Total value of company A = (price per share x number of shares) = £60 x 10 million = £600 million
Government holding = 75% of 600 million = £450 million
Disinvestment (50%) = 0.5 x 450 = £225 million
Hence £225 million is raised from the disinvestment.

Question 167: A
Total value of company B = (price per share x number of shares) = £20 x 50 million = 1,000 million
Government holding = 50% of 1,000 Million = £500 million
Disinvestment (25%) = 0.25 x 500 = £125 million

Question 168: B
Government holds 1/3 of the 30 million shares, which is 10 million shares.
It sold each share for £35, £5 less than the £40 market price. Hence the additional revenue for selling at the market price would have been £5 per share x 10 million shares = £50 million.

Question 169: B
Government holds 25% of 40 million shares, which is 10 million shares.
The price of each share fell from £30 to £35, so fell by £5 per share.
If the price of 10 million shares fell by £5, the total fall was £50 million.

Question 170: C
Government holds 12.5% of 50 million shares, which is 6.25 million.
The price has risen by £5 per share, so the total rise is £31.25 million.

Question 171: B
Total value of option A = £10 x 60 million x 75% = £450 Million
Total value of option B = £20 x 50 million x 50% = £500 Million
Hence they will fetch difference values, with B fetching more.

SET 36

Question 172: C
In 2011-2012, food grain production was 100, and this was a 25% increase on 2010-2011.
If 100 is 125%, then 100% = 80. Hence food production in tonnes in 2010-2011 was 80 tonnes.

Question 173: B
Target production (2011-12) = 60 tonnes
Actual production (2010-11) = 50/125% = 40 tonnes
Difference = 60 - 40 = 20 tonnes

Question 174: B
Difference = Target - Actual = 50 - 40 = 10 tonnes

Question 175: D
Cotton production (2010-11) = 30/120% = 25 tonnes
Jute production (2010-11) = 20/125% = 16 tonnes
Combined = 25 + 16 = 41 tonnes

Question 176: D
Food grain production (2010-11) = 100/125% = 80 tonnes
Oil seeds production (2010-11) = 50/125% = 40 tonnes
Difference = 80 - 40 = 40 tonnes

Question 177: B
Total production (2011-12) = 100+50+40+30+20 = 240 tonnes
Cotton as a percentage of total = 30/240 = 12.5%

SET 37

Question 178: D
Sales of product B in Feb = 7,000
Total sales of all products in Feb = 10,250 + 7,000 + 3,750 + 3100 = 24,100
Percentage of product B's sales = 7,000/24,100 = 29%

Question 179: D
Percentage increase:
Product A = (11,000-10,500)/10,500 = 4.76%
Product B = (7,500-7,250)/7,250 = 3.45%
Product C = (4,250-4,000)/4,000 = 6.25%
Product D = (4,000-3,500)/3,500 = 14.29%

Hence product D witnessed highest percentage growth.

However to answer this question more quickly, look at the numbers – the numbers are giving you a clue. You can visually see that product D's sales' values have gone up the equal maximum amount of £500. But it is also apparent that the absolute value of sales is the lowest, therefore you can deduce that D is the largest percentage increase without actually doing any sums!

Question 180: C
Sales of product C in May = 4,250 x 1.2 = 5,100
Therefore sales of product D in May = 5,100
Percentage increase in sales of D from April to May = (5,100 – 4,000)/4,000 = 27.5%

Question 181: D
Sale of products (A+C) in January = 13,000
Sale of products (A+C) in April = 15,250
Percentage increase in combined sale from January to April = (15,250 – 13,000)/13,000 = 17.31%

Question 182: A
Sale of product (A+B) in May = 1.2 x (11,000+75,00) = 22,200
Sale of product (C+D) in May = 1.3 x (4,250+4,000) = 10,725
Total sales in May = 32925

Question 183: D
Sale of product A in May = 1.2 x 11,000 = 13,200
Sale of product (B+C+D) in May = 1.1 x (7,500+4,250+4,000) = 173,25
Total sales in May = 13,200+17,325 = 30,525
Percentage of sales of A over total sales = 13,200/30,525 = 43.24%

SET 38

Question 184: C
Overall profit is 30% of revenue. Since profit = £2.5 million, overall revenue = 2.5/30 x 100 = £8.3 Million

Question 185: B
Let us assume cost per article = C; Total number of articles = N
Cost per consignment:
Rail = 25C/30N = 0.83C/N
Road = 25C/45N = 0.56C/N
Air = 50C/25N = 2C/N
Hence road has the lowest cost per consignment.
However if you look at the figures, a shortcut is apparent. Road occupies by far the greatest number of consignments, but the cost is the equal lowest in the business. Therefore at a glance you can see the answer is road, even before you open the calculator and start doing unnecessary sums.

Question 186: A
The ratio will be the same for any number of consignments as the proportions are preserved.
Ratio of total revenue to total cost = £20:£5 = 4:1

Question 187: C
From the table it is given that 50% of the total costs are associated with air transportation, so 50% of the total costs are due to air travel.
50% x £54,000 = £27,000

Question 188: D
From the table it is given that 30% are delivered by rail and 45% by road, so 75% of consignments are delivered by rail and road taken together.
75% x 17,145 = 12,859

SET 39

Question 189: A

This is a difficult question that would be worth "flagging for review". Set the percentage of lead in alloy A to a, and the percentage of tin in alloy C to b. We can then find the percentage of copper in each as a function of a and b.

Alloy	Zinc	Tin	Lead	Copper	Nickel
A	10%	40%	a%	(40-a)%	10%
B	25%	15%	50%	5%	5%
C	15%	b%	20%	(30-b)%	35%

The key thing here is to use the composition of Alloy G. We can find the composition of Alloy G in terms of a and b and then set the amounts of tin, lead and copper equal to each other to find a and b:

For Alloy G, the percentages will be weighted according to the proportion A:B:C = 2:1:3:
$2/6 (40) + 1/6 (15) + 3/6 (b) = 2/6 (a) + 1/6 (50) + 3/6 (20) = 2/6 (40-a) + 1/6 (5) + 3/6 (30-b)$

$80 + 15 + 3b = 2a + 50 + 60 = 80 -2a + 5 + 90 -3b$
Solving above equation, we will get values:
$95 + 3b = 2a + 110$
$2a + 110 = 175 - 2a - 3b$
$3b = 65 - 4a$
$2a = 95 - 110 + 65 - 4a$
$6a = 50$
$a = 50/6$
$b = 95/9$
Percentage of Lead in alloy A = a = 50/6% = 25/3% = 8.33%

Question 190: C

Using our solution from the previous question, we found that the percentage of Tin in alloy C, b, was:
$b = 95/9$
Percentage of Tin in alloy C = 95/9 = 10.6%

Question 191: B

Zinc percentage in alloy X is equal to the average of the percentages of the composite alloys, as they are present in equal proportions. This can be found by adding together and dividing by 3.
$X = (10+25+15)/3 = 50/3 = 16.67\%$

Question 192: D

To solve, subtract the amounts of the known metals to find the remaining metal, which is equal to the percentage of Tin and Copper combined in alloy C. We know there are no other components as this is stated in the question.

$(100\% - 15\% - 20\% - 35\%) = 30\%$

Question 193: B

We know the percentages of tin in each of the alloys which make up Alloy G, and the composition of Alloy G. Alloy G is made up of alloys A:B:C in the ratio 2:1:3. Alloy A has 40% tin, alloy B has 15% and alloy C has 95/9% tin. Hence the percentage in Alloy G is (2/6 x 40)+(1/6 x 15)+(3/6 x 95/9) = 21.11%.

Question 194: C
Percentage of elements in alloy G:

We know that Alloy G has 21.11% Tin from the last question. We also know from the initial explanation that it has the same concentration of tin, lead and copper. This is 3 elements with the same concentration. We then need to work out how much nickel and zinc there is to check whether there is a 4th.

Alloy G is made up of alloys A:B:C in the ratio 2:1:3. Alloy A has 10% nickel, alloy B has 5% and alloy C has 35% nickel. Hence the percentage in Alloy G is ((2x10)+(1x5)+(3x35))/6, = 21.6666.

We can also work out from the fact that these 4 elements plus Zinc are 100% of the total that Zinc is 15%.

Zinc = 15%
Tin = 21.11%
Lead = 21.11%
Copper = 21.11%
Nickel = 21.67%

SET 40

Question 195: A
Number of persons who voted in favour of Hilary Clinton = 60% of 17% of 11,500 = 0.6x0.17x11,500 = 1173.

Question 196: D
We cannot find the number of people living in New York, as we do not know the proportion of citizens of New York who voted for people other than Robert Guiliani. We can only say that there was a minimum of 460 New York citizens. We cannot determine the actual number.

Question 197: D
0.39 x 11,500 = 44,85 votes for Bush.
Therefore 4,485/(11,500 x 0.8) gives the proportion of US citizens who voted for Bush.
This equals 48.8%.

Question 198: C
10% of 40% of people surveyed are in favour of Rumsfield and employees of the federal government, so 10% of 40% of 11,500 = 460 people.
Also from the table, Rumsfield has 5% of 11500 votes = 575 people who are in favour of him in total.
Hence the number of people who are in favour of Rumsfeld who are not employees of the federal government is 575-460 = 115.

Question 199: C
The number of people polled was constant. The decrease in percentage is from 16% to 2%, i.e. 14 % of 11,500 = 1,610 people.

Question 200: A
The number of people polled was constant. The decrease in percentage is from 40% to 39%, i.e. 1% of 11,500 = 115 people.

SET 41

Question 201: E
15% indicates that only 85% of the total price is to be paid.
50p is the price per unit for premium mugs for quantity below 49 units.
20p is the price per unit for basic mugs for quantity below 49 units.
5p is the price per unit for logo on basic mugs for quantity below 49 units.
Using this information, we can work out:
$0.85x(5x50p+4x(20p+5p)) = 297.5p = £2.975$ rounded up to £2.98
Therefore the company will have to pay £2.98

Question 202: B
20p is the price per unit for premium mugs for quantity above 500 units.
10p is the price per unit for basic mugs for quantity between 100 and 499 units.
3p is the price to add logos to those mugs.
4p is the price per unit for black and white flyers for quantity between 10 and 99 units.
Combining this information together gives the equation:
$750x20p +130x(10p+3p)+80x4p=17010p = £170.10$ rounded down to £170
So, the company will have to pay £170.

Question 203: C
Step I.
£250 equals 25,000p
$25,000p – 50x10p = 24,500p$
10p is the price per unit for colour flyer between 10 and 99 units.
Step II. $24,500p/(10p+2p)=2041.667$. Therefore she can buy 2041 medium mugs.
Once you bought the flyers you are left with 24,500p and you want to know how many medium mugs with logo you can buy.
10p is the price per unit for medium mugs above 500 units.
2p is the price per unit for logo on medium mugs above 500 units.

Question 204: D
Step I.
2014: $70x(26p+4p)+150x5p= 2850p$
26p is the price per unit for medium mugs below 99 units in 2014.
4p is the price per unit for logo on medium mugs below 99 units in 2014.
5p is the price per unit for colour flyers above 100 units in 2014.
Step II.
2015: $70x(26p+3p)+150x4p=2630p$
26p is the price per unit for medium mugs below 99 units in 2015.
3p is the price per unit for logo on medium mugs below 99 units in 2015.
4p is the price per unit for colour flyers above 100 units in 2015.
Step III.
$1-(2630/2850)=0.077 = 8\%$

Question 205: C
$£325,750/(1.012^{\wedge}(3)) =£314,298.9$ rounded down to £314,299
1.2% equals 0.0012
3 years between 2012 and 2015.

SET 42

Question 206: B

The easiest way to solve this problem is using a Venn diagram. The Venn diagram below shows all possible combinations of the three devices each student can have as well as the number of students with a combination of devices. The sum of the numbers in the Venn diagram must be equal to the total number of students.

Using the information given in the graphs, we know there are:

- 30 students with all three devices
- 50 students with smartphone only
- 40 students with tablet only
- 50 students with laptop only
- 180 students with smartphone
- 190 students with tablet
- 200 students with laptop

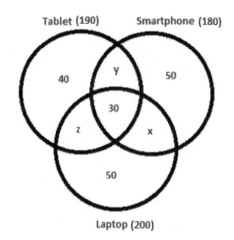

Therefore we can construct the following Venn diagram:

Laptop: $200=50+30+z+x \rightarrow z=120-x$
Smartphone: $180 =50+30+x+y \rightarrow y=100-x$
Tablet: $190=40+30+y+z \rightarrow y=120-z$

So, we see that $y=100-x$ and $y=120-z$ and thus $z=20+x$
Then, we see that $z=20+x$ and $z=120-x$ and thus $x=50$
Plug it in to see, $z=20+x=70$ and $y=100-x=50$

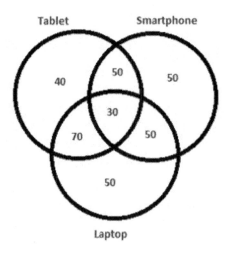

Now, the total number of students is the sum of the numbers in the Venn diagram.

Question 207: C

Using the previously constructed Venn diagram, we can see the total number of students with a smartphone and a laptop is 50.

Question 208: A

180 students have smartphone
100 students have both tablet and laptop
So, $180-100=80$, 80 students.

Question 209: C

Total number of students: 340
80 students have both smartphone and laptop.
$80/340=0.25 \rightarrow 23.5\%$

Question 210: C

185/345=0.536 → 54%

185 – no of students with smartphone (AFTER)

345 – total no of students (AFTER)

BEFORE   **AFTER**

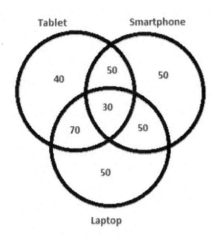   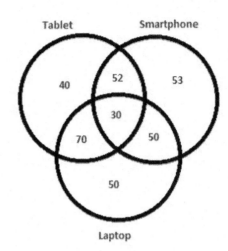

SET 43

Question 211: B

From 7.10am, trains run in every 20 minutes and hence the first one after 2pm will depart at 2.10pm - in every 20 minutes. This gives us three trains per hour from 2.10pm which would come to 12 trains before 6.15pm.

Question 212: C

From 8.18am, trains run in every 20 minutes and hence the first one Antonia can catch is the 2.58pm train. As it takes 21 minutes to get to London Liverpool Street station, she can arrive by 3.19pm.

Question 213: D

There are three trains in the morning before 7.10am. Thereafter, 3 trains per hour and the last one departs at 23.10pm. There are 16 hours between 7.10am and 23.10pm, so 3+16x3=51. 51 trains run all day.

Question 214: B

The first train after 3pm departs at 3.10pm from Cambridge Station and arrives at Tottenham Hale by 4.28pm.

Question 215: D

Using the equation: velocity = distance/time

7 minutes = 7/60 hour from Audley End to Bishops Stortford

10.5 miles between Audley End and Bishops Stortford

10.5 miles / (7/60) hour = 90 miles per hour

SET 44

Question 216: A

The top five rivers in the table are the longest.

Take the length in miles from column 2 and add them together.

Sum: 220+215+185+143+134=897 miles

Question 217: B

Firstly, you add the drainage area of all seven rivers from column C. Then you add the increase in drainage area of River Thames (1345.5 square mile). Finally, you can calculate the percentage increase, dividing New by Old.

Old: 4,409+4,994+4,029+3,236+1,597+560+431=19256

New: 1345.5+4,409+4,994+4,029+3,236+1,597+560+431=20601.5

Percentage: 20601.5/19256=1.06987 → 7%

Question 218: C

River Thames: 4,994 square mile

River Wye: 1,597 square mile

Percentage:

4,994/1,597=3.127 → 313%

Question 219: A

Lowest average discharge: River Tay – length: 117 miles

Highest average discharge: River Trent – length: 185 miles

Difference: 185-117=68

SET 45

Question 220: B

Remember that only the SIX longest rivers shall be included in the calculation.

Use the figures from column 3 to obtain the equation:

(4,409+4,994+4,029+3,236+1,597+560)/6=3137.5

Question 21: C

The largest section on the pie chart for Scottish GDP is Government and Other Services – it accounts for 29% of total Scottish GDP.

Question 222: B

Using the figures we are given: 4,2tn pounds and 5% (1.05) growth, we can plug them into the equation: GDP_{2014} x 1.05 = 4.2tn

This can be arranged as: → GDP_{2014}=4,2/1,05=4

Question 223: E

This requires a simple subtraction.

Manufacturing in Scottish GDP 14%

Business Services and Finance in UK GDP 28%

Difference: 28-14=14 → 14%

Question 224: B

We can use the figures given us to formulate the following equation:

£1.2tn (GDP in 2014) x 1.03 (3% GDP increase) x 0.01 (1% share of Agriculture, Forestry and Fishing) = 0.01236tn

Question 225: D

The three largest sections of the UK GDP pie chart are:

Distribution, Hotels and Catering 15%

Business Services Finance 28%

Government and Other Services 24%

SET 46

Question 226: C
The most-watched TV show is Geordie shore
Year 1 divided by Year 2
9.2/12.5=1.35869 → 36%

Question 227: E
The Voice: 3.4 and 4.1 in year 1 and 2 respectively
Britain's got talent: 5.2 and 5.6 in year 1 and 2 respectively
Year I: 5.2+3.4=8.6
Year II: 5.6+4.1=9.7
Difference: 9.7-8.6=1.1

Question 228: C
The show that the fewest percentage of males watched in year 2 was The Voice.

Question 229: B
Year 2 (all five TV show) / year1 (all five TV show) = (4.1+5.6+1.4+12.5+3)/(5+9.2+1.3+5.2+3.4) = 1.1037 →
10%

Question 230: D
Total population of females in year 2: 30 million
Number of females who watched Geordie Shore: 30x0.21=6.3 million
Number of males who watched Geordie Shore: 12.5-6.3=6.2 million

SET 47

Question 231: D
Basic plan: £45.29 + £7.50 = £52.79
250 free texts – more than 75 free minutes: (300-250) x15 = £7.50
Premium plan: £47.89
More than 300 free texts – more than 75 free minutes
£52.79 - £47.89 = £4.90

Question 232: C
Premium plan: £47.89
£60 - £47.89 = 12.11, so he has £12.11 left
Cheapest calls: Standard Calls – 6p
12.11/0.06= 201.83=201 minutes.

Question 233: E
45 texts free, 125 mins same network free, 100 minutes free with other network → 225 mins other network in
both options. Before → 225 x 0.15 = 33.75
After → 225 x 0.16 = 36. Difference is £2.25

Question 234: D
Basic plan: £45.29 + £58.50 = £103.79
(800-150) x9p= £58.50 – 150 minutes for free
Premium plan: £47.89 + £45.00 =£92.89
(800-500) x15p= 4500p = £45.00 – 500 minutes for free
Difference:
£103.79- £92.89 = £10.90

Question 235: C
Basic plan: £45.29x1.15 + (450-250)x £0.15 = £52.08+£30.00=£82.08
Premium plan: £47.89x1.15=£55.07
Difference: £82.08 - £55.07=£27.01

SET 48

Question 236: B
Davos: 15cm, 15cm, 15cm and 10cm in November, December, January, February respectively
Chamonix Mont-Blanc: 5cm, 40cm, 15cm, 20cm in November, December, January, February respectively
We can use these figures to create the equation:
(15+15+15+10)+(5+40+15+20)/8=16.875 cm

Question 237: C
Davos: 15cm, 15cm, 15cm and 10cm in November, December, January, February respectively
Average snowfall in Davos: (15+15+15+10)/4 = 13.75
Chamonix Mont-Blanc: 5cm, 40cm, 15cm, 20cm in November, December, January, February respectively
Average snowfall in Chamonix Mont-Blanc: (5+40+15+20)/4 =20
Cortina d' Ampezzo: 50cm, 50cm, 40cm and 5cm in November, December, January, February respectively
Average snowfall in Cortina d' Ampezzo: (50+50+40+5)/4=36.25
Garmisch Partenkirchen: 10cm, 15cm, 35cm, 20cm in November, December, January, February respectively
Average snowfall in Garmisch Partenkirchen: (10+15+35+20)/4=20
So the highest average snowfall was in Cotine d'Ampezzo.

Question 238: C
Inserting figures from all of the places into an equation:
December/February = (15+15+50+40)/(35+10+40+20)=120/105=1.14

Question 239: B
We are told that in November 2014 there is 30cm of snowfall (in all four areas)
To work out November 2015, we need to add all the areas: - Garmisch-Partenkirchen: 10cm, Davos: 15cm, Cortina d'Ampezzo: 5cm, Chamonix Mont-Blanc: 5cm
This gives us 35cm for November 2015
This gives us the equation: 35/30=1.1667

Question 240: D
We need to add the values for November and February for both of these places.
Cortina d' Ampezzo 5 + 40 = 45
Garmisch Partenkirchen 35 + 10 = 45
Sum: 45 + 45 = 90

SET 49

Question 241: A
Remember that two expenses relate to HSBC:

Breakfast with client	**£35.90**
Mileage to and from presentation in Cambridge	£16.80

Sum: £35.90+£16.80=£52.70
Check: Sum is greater than any other item on the list.

Question 242: B
Add all of the expenses from each company together.
£35.90+£16.80+£15.40+£20.00+£49.50=£137.60

Question 243: E

Soros Fund Management	Single ticket - Train journey to meeting in London	£15.40	Train
HSBC	Breakfast with client	£35.90	Meal
Black Rock	Return ticket – Train journey to meeting in London	£20.00	Train
MKB	Lunch with client	£49.50	Meal

Meal/Train: (£35.90+£49.50)/(£15.40+£20.00)-1=1.413 → 141%

Question 244: B
Mileage paid at £0.25 per mile for the first 100 miles each month and £0.10 thereafter
Paid: £16.80
100 miles x £0.25 per mile = £25 → this is higher than £16.80 that Kevin actually paid, so the £0.25 per mile applies. → £16.80/£0.25=67.2

Question 245: D

Mileage to and from presentation in Cambridge	£16.80	Travel
Single ticket - Train journey to meeting in London	£15.40	Travel
Breakfast with client	£35.90	Meal
Return ticket – Train journey to meeting in London	£20.00	Travel
Lunch with client	£49.50	Meal

The total sum of travel expenses - £16.80+£15.40+£20.00=£52.2
75% of travel expenses - £52.2 x 0.75 =£39.15
Difference - £52.2-£39.15= £13.05
Accommodation and meal expenses - £35.90+£49.50=£85.40
90% of accommodation and meal expenses - £85.40x0.9=£76.86
Difference - £85.40--£76.86=£8.54
Sum of differences: £13.05+£8.54=£21.59

SET 50

Question 246: A
£30 per hour and Total cost £50 → £50-£30=£20

Question 247: B
Gamma for 6 hours = £25 deposit + (£115 x 6 hours) = 25 + 690 = £715
Beta for 6 hours = £20 deposit + (£90 x 6 hours) = 20 + 540 = £560
Difference = 715 – 560 = **£155**

Question 248: B
Using the information from the table:
Alpha: £50
Gamma: £140
Beta: £110
Delta: £250
Aron can afford to rent Alpha, Beta and Gamma for one hour (£300).
Alpha, Beta, Gamma and Delta would cost £550 which is out of his budget.

Question 249: B
Alpha: £30 per half an hour → £90 for three hours plus deposit of £20 → £110
Gamma: 3hours x £115 (cost per hour) + £25 (deposit) = £370
Sum: £110+£370=£480

Question 250: C
Old: Deposit £100 + (£150x8) = £1300
New: (£100 x 1.05) + (£150x8) = £1305

SET 51

Question 251: C
Remember to convert all of the prices into £
Around the World offers a price of 400 €.
Every 1.5€ equals the value of 1£. 400/1.5 = 266,67 rounded up to £267.
This is cheaper than the other options available.

Question 252: B
The differences from the European price in pounds are 190, 175, 180 and 200.
The average of the differences is (190+175+180+200)/4 = 186.25 rounded up to 186.3

Question 253: A
The discounted price of Take Me There is 560-50=510$, in pounds 510/2=255
This discounted ticket is £5 more expensive than the original price of In The Air (£250).
In percentages, this difference is 5/250x100 = 2%

Question 254: C
The total revenue of Around The World is 10x450+5x300=€6000, in pounds 6000/1.5=£4000
The total revenue of Good Fly is 25x100+12x300= £6100
The difference between revenues is £2100, which is 2100/6100x100=34.42....% of the higher revenue, rounded up to 34.43%

Question 255: C
The original price in pounds is 610/2=£305, the new price is 610/2.5=£244.
The difference is 305-244=£61

SET 52

Question 256: D
Looking at the pie chart for Indian GDP, the largest section is for manufacturing, which accounts for 24% of total Indian GDP.

Question 257: C
UK: Government and Other Services (24%) / Business Services and Finance (28%) = 86%
India: Government and Other Services (9%) / Business Services and Finance (15%) = 60%
UK – India = 26%

Question 258: A
The three largest sections of the pie chart for Indian GDP are:
Distribution, Hotels and Catering 20%
Business Services Finance 15%
Manufacturing: 24%

Question 259: D
Bottom two performing sector in the UK: Agriculture, Forestry and Fishing (1%), Other Production (4%)
Contribution: £4.2tn (total GDP) x 0.05 (1%+4%) = £210Bn
Bottom two performing sector in India: Other Production (3%), Construction (6%)
Contribution: £2tn (total GDP) x 0.12 (9%+3%) = £180Bn
UK minus India: £210Bn - £180Bn = £30Bn

Question 260: D
Largest sector in the UK: Business Services and Finance: £4.2tn x 0.28 =£1.176tn
Largest sector in India: Manufacturing: £2tn x 0.2 =£400Bn
Size of the UK economy: £4.2tn
Size of Business Services and Finance in the UK is £1.176tn, which is the largest

SET 53

Question 261: A
1500 JPY at 1 JPY = 0.013 USD → 1500 x 0.013 = 19.50

Question 262: D
I JPY = 0.022 GBP
1 JPY = 0.015 USD
GBP/USD = 0.022 / 0.015 = 1.47

Question 263: A
% increase: ((JPY/CAD) 2012 / (JPY/CAD) 2011) - 1 = 0.119 / 0.115 = 0.0348. → 3.5%.

Question 264: A
USD 1300 at USD/JPY rate of 0.010 in 2010 → JPY 130,000
USD 1300 at USD/JPY rate of 0.015 in 2013 → JPY 86,667
Difference: JPY 130,000 - JPY 86,667 = 43,333

Question 265: C
Looking at the table it is easy to notice that EUR exchange rate is near constant over the given time period except for one change in 2011 by 0.001. This is smaller than all the other currency changes

	2010	2011	2012	2013
GBP	0.021	0.020	0.019	0.022
USD	0.010	0.013	0.013	0.015
EUR	0.015	0.016	0.015	0.015
CAD	0.123	0.115	0.119	0.125

SET 54

Question 266: D
Boston: 4.4 – 4 = 0.4 million
Chicago: 3.3 – 3.5 = -0.2 million
Denver: 2.1 – 2 = 0.1 million
El Monte 1.5 – 0.8 = 0.7 million
El Monte has the highest growth in population

Question 267: A
Boston: 542,000– 235,675 = 306,325
Chicago: 350,685 – 345,526 = 5159
Denver: 249,990 – 231,456 = 18534
El Monte 62,044 – 54,000 = 8044
Question 268: D

Boston: 542,000/4.4 million = 0.123
Chicago: 350,685/3.3 million = 0.106
Denver: 249,990/2.1 million = 0.119
El Monte 62,044/1.5 million = 0.041

Question 269: A
2014:
Boston: 542,000/4.4 million = 0.123
Chicago: 350,685/3.3 million = 0.106
Denver: 249,990/2.1 million = 0.119
El Monte 62,044/1.5 million = 0.041
2009:
Boston: 235,675/4 million = 0.059
Chicago: 345,526/3.5 million = 0.099
Denver: 231,456/2 million = 0.116
El Monte 54,000/0.8 million = 0.068
Difference:
Boston: 0.123 - 0.059 = 0.064
Chicago: 0.106 - 0.099 = 0.007
Denver: 0.119 - 0.116 = 0.003
El Monte: 0.041 - 0.068 = -0.027

Question 270: D
2009: Boston: 235,675/4 million = 0.059
2014: Boston: 542,000/4.4 million = 0.123
Difference: Boston: 0.123 - 0.059 = 0.064

SET 55
Question 271: E
Starting rate: 15 % of 2,450 → £367.5
Basic rate: 25% of 2,450 – 33,500 so 25% of £31,050 → £7762.50
Higher rate: 40% of 33,500 – 37,000 → 40% of £3500 → £1400
Overall: £367.5 + £7762.25 + £1400 =£9530

Question 272: A
They earn the same salary.

Question 273: D
Starting rate: 15 % of 2,450 → £367.5
Basic rate: 25% of 2,451 – 33,500 so 25% of £31,049 → £7762.25
Higher rate: 40% of 33,500 – 42,000 → 40% of £8500 → £3400
Monthly average: (£367.5+ £7762.25+ £3400)/12 =£960.80

Question 274: B

Adam's income tax with £37,000 salary

Starting rate: 15 % of 2,450 → £367.5

Basic rate: 25% of 2,451 – 33,500 so 25% of £31,049 → £7762.25

Higher rate: 40% of 33,500 – 37,000 → 40% of £3500 → £1400

Overall: £367.5+ £7762.25+ £1400 =£9529.75

Adam's income tax with £41,000 salary

Starting rate: 15 % of 2,450 → £367.5

Basic rate: 25% of 2,451 – 33,500 so 25% of £31,049 → £7762.25

Higher rate: 40% of 33,500 – 41,000 → 40% of £7500 → £3000

Overall: £367.5+ £7762.25+ £3000=£11129.75

Difference: £11129.75 - £9529.75= £1600

Question 275: E

2014-15: £2,450

2015-16: £2,730

Difference: £2,730 - £2,450 = £280 increase

SET 56

Question 276: D

Follows from the table.

Question 277: B

£654,150,000 (total sales) / 7 (number of projects) = 93,450,000

Question 278: E

£45,000,000 x 1% + £180,000,000 x 4% = 450,000 + 7,200,000 = 7,650,000

Question 279: A

Follows from the table.

Question 280: A

Sum: £45,000,000 + £150,453,000 + £180,000,000 +£654,150,000 = £1,029,603,000

SET 57

Question 281: C

Previous day share price x (1 + change from previous day) = current share price

Change from previous day is +0.13%

Current share price is £134.432

£134.432 / (1-0.0013) = £134.257

Question 282: C

British Land highest: 54.934

British Land lowest: 54.914

Difference: 54.934 - 54.914 = 0.02

Question 283: E
Calculate previous day share price first, then calculate market capitalisation
HSBC current share price: 25.432
HSBC volume: 7,345,321
Market capitalisation: 25.432 x 7,345,321 = 186,806,203.672
Change from previous day: -0.03%
25.432/(1-0.0003) = 25.4396318896
Therefore, the market capitalisation previous day was 186,806,203.672 / (1 + 0.03) =186,750,178.618
25.439 etc. x 7,345,321 = 186, 862, 262.351

Question 284: C
Follows from the table.

Question 285: A
BP market cap: 286.123 (share price) x 4,431,748 (volume) = 1,268,025,033.004
British Land market cap: 54.923 (share price) x 4,999,432 (volume) = 274,583,803,736
Difference: 1,268,025,033.004 - 274,583,803,736 = 993,441,229.268

SET 58

Question 286: D
Psychology: 10/6 = 1.66
To determine the course with a similar ratio of men to women as psychology, we need to calculate the ratios for all of the other courses
Mathematics: 8/7 =1.14
Physics: 10/15 = 0.66
Programming: 4/5 = 0.8
Literature: 12/8 = 1.5
History: 7/7 =1
From these figures, we can see that Literature has the most similar ratio.

Question 287: C
Physics: 10/15 = 0.66
Similarly to the previous question, we need to calculate the ratios for all other courses
Psychology: 10/6 = 1.66
Mathematics: 8/7 =1.14
Programming: 4/5 = 0.8
Literature: 12/8 = 1.5
History: 7/7 =1

Question 288: E
Can't tell because students can take more than one course.

Question 289: E
As one student can take more than one course and we do not have any information about the total number of students we cannot say.

Question 290: C
Before ratio: 10:6 = (10/6):1 = 1.66:1
After ratio: 13:6 = (13/6):1 = 2.16:1
Difference: 2.16-1.66 = 0.5

SET 59

Question 291: A
Women in employment: 85.5% (employed) + 5% (self-employed) + 2.7% (employed other) = 93.2%
Men in employment: 81.3% (employed) + 8% (self-employed) + 4.5% (employed other) = 93.8%
Difference: 93.8% - 93.2% = 0.6

Question 292: D
Women self-employed: 5% - Women unemployed: 6.8% → sum: 11.8%
Men self-employed: 8% - Men unemployed: 6.1% → sum: 14.1%
Ratio men to women: 14.1% / 11.8% = 1.2

Question 293: E
21 million (women employed) x 0.05 (5% self-employed) = 1.05 million
If there are 21 million women in employment → 21 million are either employed, self-employed or employed other – so 21 million women is (85.5+5+2.7) 93.2% of the country. This means 6.8% of the country is 1532189 so country population is 22,532,189. 5% of this is 1,126,609 so answer should be 1.13 million

Question 294: A
Men unemployed: 6.1% → 31 million x 0.061 = 1.891
Women unemployed: 6.8% → 32 million x 0.068 = 2.176
Difference 2.176 – 1.891 = 0.285

Question 295: D
Women in employment (as defined above): 93.2
Unemployed women: 6.8
93.2/6.8 = 13.7

SET 60

Question 296: C
Energy cost: electricity plus gas. Note that Gas and Electricity together add up to ¼ of the total area. And therefore, 360 pounds x ¼ = 90 pounds

Question 297: D
Household purchases, travel and gas costs – ½ of the total area → 180 pounds
Energy costs – ¼ of the total area → 90 pounds
Ratio: 180 / 90 = 2

Question 298: B
The largest section of the pie chart is for 'Travel'

Question 299: D
Household purchases, travel and gas costs – ½ of the total area → 180 pounds
Energy costs – ¼ of the total area → 90 pounds

Question 300: E
The smallest section of the pie chart is for 'TV subscription'

Abstract Reasoning Answers

Q	A	Q	A	Q	A	Q	A	Q	A	Q	A
1	B	51	Neither	101	A	151	A	201	C	251	A
2	Neither	52	B	102	A	152	Neither	202	A	252	D
3	A	53	A	103	Neither	153	A	203	A	253	D
4	A	54	Neither	104	B	154	B	204	D	254	D
5	Neither	55	B	105	Neither	155	A	205	D	255	C
6	B	56	A	106	A	156	Neither	206	A	256	B
7	A	57	Neither	107	Neither	157	A	207	C	257	C
8	A	58	B	108	A	158	A	208	B	258	C
9	Neither	59	Neither	109	B	159	Neither	209	B	259	A
10	Neither	60	A	110	B	160	B	210	A	260	B
11	A	61	A	111	B	161	B	211	D	261	B
12	B	62	A	112	Neither	162	A	212	A	262	C
13	Neither	63	B	113	Neither	163	A	213	C	263	A
14	A	64	B	114	B	164	B	214	A	264	A
15	A	65	A	115	A	165	A	215	D	265	B
16	B	66	A	116	A	166	A	216	D	266	A
17	B	67	A	117	A	167	B	217	A	267	C
18	A	68	B	118	Neither	168	Neither	218	C	268	A
19	A	69	Neither	119	B	169	A	219	D	269	A
20	A	70	Neither	120	Neither	170	B	220	C	270	D
21	A	71	B	121	Neither	171	Neither	221	A	271	A
22	A	72	A	122	Neither	172	B	222	D	272	C
23	B	73	Neither	123	A	173	B	223	D	273	C
24	B	74	Neither	124	A	174	A	224	B	274	C
25	Neither	75	A	125	B	175	Neither	225	C	275	D
26	B	76	A	126	Neither	176	B	226	B	276	D
27	A	77	A	127	A	177	Neither	227	D	277	A
28	A	78	B	128	Neither	178	B	228	C	278	B
29	B	79	A	129	Neither	179	B	229	A	279	C
30	B	80	Neither	130	B	180	A	230	A	280	B
31	B	81	B	131	Neither	181	Neither	231	A	281	B
32	Neither	82	Neither	132	Neither	182	Neither	232	D	282	A
33	Neither	83	Neither	133	Neither	183	A	233	A	283	A
34	A	84	A	134	A	184	A	234	C	284	D
35	A	85	A	135	B	185	Neither	235	B	285	B
36	B	86	B	136	A	186	Neither	236	C	286	C
37	A	87	A	137	B	187	Neither	237	C	287	C
38	Neither	88	B	138	Neither	188	A	238	C	288	A
39	A	89	Neither	139	Neither	189	B	239	D	289	C
40	Neither	90	B	140	B	190	A	240	A	290	C
41	B	91	Neither	141	A	191	Neither	241	A	291	D
42	Neither	92	A	142	Neither	192	A	242	D	292	A
43	A	93	Neither	143	B	193	Neither	243	B	293	A
44	B	94	A	144	A	194	B	244	A	294	C
45	Neither	95	A	145	B	195	Neither	245	A	295	B
46	B	96	B	146	Neither	196	B	246	D	296	D
47	Neither	97	B	147	B	197	Neither	247	C	297	A
48	A	98	A	148	Neither	198	B	248	C	298	C
49	B	99	B	149	A	199	Neither	249	D	299	A
50	Neither	100	B	150	A	200	B	250	B	300	A

	Set A Rule	**Set B Rule**
Set 1	**3** Shapes are white **2** Shapes are black	**2** Shapes are white **3** Shapes are black
Set 2	**2** Shapes are white **1** Shape is black	**1** Shape is white **2** Shapes are black
Set 3	There is always a **triangle** in the **top left corner**	There is always a **quadrilateral** in the **bottom right corner**
Set 4	There are an **even** number of rectangles	There are an **odd** number of rectangles
Set 5	Each circle has at least one tangential line.	At least one circle is **intersected** by a line.
Set 6	**None** of the shapes intersect with each other.	At least **one** of the shapes intersects with another shape/line.
Set 7	The total number of dots is **10**	The total number of dots is **9**
Set 8	The four-sided shapes are in **different** section of the box.	The four-sided shapes are in the **same** section of the box.
Set 9	All boxes have a **five-pointed star** in the **Centre**	All boxes have a **triangle** in the **bottom left corner**
Set 10	The black **circle** is always in the **bottom left corner**	The two small **dots** are always in the **top right corner**
Set 11	The circles **are all tangential to** each other	The circles **intersect** each other
Set 12	There are always **three squares, two triangles** and one circle	There are always **three triangles, two squares** and one circle
Set 13	The sum of the edges is **odd**	The sum of the edges of is **even**
Set 14	The sum of the edges is **twelve**	The sum of the edges is **fourteen**
Set 15	Each box contains **4** right angles	Each box contains **6** right angles
Set 16	Each black shape has a **larger** white counterpart	Each black shape has a **smaller** white counterpart
Set 17	Number of Dots = Number of **Stars**	Number of Dots = Number of **Diamonds**
Set 18	Even-sided polygons are **black** Odd-sided polygons are **white**	Even-sided polygons are **white** Odd-sided polygons are **black**
Set 19	Number of Stars = Number of **Triangles**	Number of Stars = Number of **Diamonds**
Set 20	There are **no** right angles	There is at least **one** right angle

Set 21	There is a **horizontal** line for each rectangle and a **vertical** line for each square	There is a **vertical** line for each rectangle and a **horizontal** line for each square
Set 22	One shape has **2** more edges than the other	One shape has **3** more edges than the other
Set 23	Even-sided polygons are **black** Odd-sided polygons are **white**	Even-sided polygons are **white** Odd-sided polygons are **black**
Set 24	The number of triangles + circles = **3**	The number of triangles + circles = **2**
Set 25	The vector sum of the arrows is **left/right**	The vector sum of the arrows is **up/down**
Set 26	There are **nine** corners in total	There are **ten** corners in total
Set 27	There is one **more** shape outside the circle than inside the circle	There is one **less** shape outside the circle than inside the circle
Set 28	The arrows only form **acute** angles	The arrows only form **obtuse** angles
Set 29	Shapes are made up of arrows all pointing at the same corner	Shapes are made up of arrows all pointing at different corners
Set 30	A central shape is intersected by an **even** number of other shapes	A central shape is intersected by an **odd** number of other shapes
Set 31	There is an **odd** number of corners in total	There is an **even** number of corners in total
Set 32	All **white** shapes are located on the left side	All **black** shapes are located on the left side
Set 33	The number of edges is a prime number	The number of edges is **not** prime number
Set 34	There are an even number of **black** shapes	There are an even number of **white** shapes
Set 35	All shapes contain right angles	All shapes **don't** contain any right angles
Set 36	Every shape is a quadrilateral	Every shape is **not** a quadrilateral
Set 37	There is always a **circle** in the top left hand corner	There is always a **triangle** in the top left hand corner
Set 38	**Two** corners are occupied by shapes	**Three** corners are occupied by shapes
Set 39	There is only **one** black shape	There are only **two** black shapes
Set 40	There are always three members of the same colour and shape, **but** different sizes	There are always three members of the same colour, shape **and** size

Question 201: C
As the sequences progresses, the number of objects within each frame increases by two.

Question 202: A
The total number of objects in each panel totals seven. In the bottom half, the number of white circles increases by one each panel. In the top half, the number of black circles alternates between one and three between each panel with the grey circles making up the remainder to total seven.

Question 203: A
The sequence alternates between a 3-edged and 4-edged zig-zag. Within each zig-zag will be a black circle and a grey shape.

Question 204: D
The circle and triangle begin at the bottom-left corner and rotate clockwise throughout the panels and cycle between the colours white-grey-black. Each panel also contains two white stars and a square.

Question 205: D
The top-right and bottom-left shapes are both the same colour and alternate between white and black. The top-left and bottom-right shapes are also both the same colour and alternate between black and white.

Question 206: A
Whilst the curve rotates 180° with each progressing frame, the total number of sides increases by 1. The colours are distracting factors.

Question 207: C
As the sequence progresses, the difference between the number of squares and circles increases by one.

Question 208: B
The shapes within the smaller white squares rotate 90° clockwise within each frame. The total number of white shapes is greater than the total number of black shapes by one.

Question 209: B
The total number of right-angles in each frame increases by one.

Question 210: A
There are a number of distractors in this series – pay attention to only the arrows. Each frame has an arrow that rotates in both the direction it faces (up-right-down-left) and the quadrant in which it is located. The arrow alternates between the colours white and black.

Question 211: D
The total number of sides for white shapes increases by one each frame. This is the same as the number of black shapes

Question 212: A
Each of the circles in the top row moves to the left once, each of the circles in the bottom row move to the right once. The circles in the middle row flip in the horizontal axis.

Question 213: C
The total number of edges increases by five each time.

Question 214: A
As the sequence progresses, the number of black circles subtract the number of white circles increases by two (6-8-10-12-14).

Question 215: D
The five-pointed start rotates clockwise by 90° in each frame. One of these points will face either up-down-left-right. If the arrow is pointing in the same direction as this, it will be coloured black.

Question 216: D
The total number of sides of all the shapes increases in the sequence 6, 9, 12, 15... When the total number of sides is odd the shapes are black, when the total is even the shapes are white.

Question 217: A
The number of shapes with right-angles increases by one.

Question 218: C
As the sequence progresses, the quadrant in which the black shape sits rotates. The two closest shapes are always grey and the furthest shape is always white.

Question 219: D
Each shape is rotated 90° clockwise, however their position is random.

Question 220: C
The number of arrows pointing directly at another shape alternates between one and three.

Question 221: A
All circles move closer towards each other and the circles in the bottom row cycle between white-grey-black. The other two circles are the same colour as the circle in their column.

Question 222: D
Each frame contains a two or three-sided 'container' that rotates. Shapes with right angles can only be placed within the container whilst shapes without right angles can only be placed outside the container.

Question 223: D
The difference between the number of the most abundant shape subtract the least abundant increases by one each frame.

Question 224: B
As the sequence progresses, the number of edges outside the central square increases by the number of edges within the central square. The colour of all the shapes outside the central square is the same as that of the shape within the central square in the previous frame.

Question 225: C
Every square alternates its colour between white and grey. New squares that are added are grey.

Question 226: B
Each shape is replaced by another shape with one more edge. Instead of a six-sided shape, two triangles are created instead. If there is another shape with the same number of edges, the shape is coloured black.

Question 227: D
As the sequence progresses, an increasing number of circles are added to the frame. Every circle cycles between the colours white-grey-black.

Question 228: C
Every arrow splits into two arrows pointing in opposite directions, the whole frame then rotates 90°. After pointing in both the up/down and left/right direction the arrows change colour.

Question 229: A
As the sequence progresses, every shape is moved to a random location in each subsequent frame. If the shape was intersecting another shape, it alternates colour between white and grey. All collisions are coloured black.

Question 230: A
The number of stars in the top-right corner increases by one and alternates colour between black-grey-white. The circle-triangle pair rotates clockwise 90° but remains in the bottom left corner and alternates between colours white-grey-black.

Question 231: A

Every square/triangle is split into two right-angled triangles. The circles alternate between being white and outside the rectangle, and being black and inside the rectangle.

Question 232: D

The entire frame rotates by 90° clockwise. The triangles alternate between white and grey.

Question 233: A

The central grey shape alternates between a four-point and six-point star. The total number of other shapes will be the same as the number of points on the star. All arrows are coloured black.

Question 234: C

The overall pattern flips vertically every frame. As the pattern progresses, two more triangles are coloured in.

Question 235: B

The total number of circles increases by one each frame. All circles are coloured grey.

Question 236: C

The total number of intersections that are coloured black increases by one each frame. The colours of the large circle alternate between white and grey.

Question 237: C

The arrows in each frame point in different directions. As the series progresses, the direction that is missed rotates between left-up-right-down

Question 238: C

The pentagon rotates 90° anti-clockwise and alternates between white and grey.

Question 239: D

The black central shape alternates between a square and a pentagon. The number of triangles present in the frame is equal to the number of sides of the central shape.

Question 240: A

Black shapes rotate clockwise 90° whilst white shapes rotate anti-clockwise 90° each frame.

Question 241: A

Every circle moves one space to the right.

Question 242: D

The number of circles in the top half increases by one whilst the number in the bottom half decreases by one. The colour of the circles within each half alternates between white and black.

Question 243: B

The triangles in the outer section rotate anti-clockwise every frame whilst the central diamond rotates clockwise.

Question 244: A

The colours in the central diamond alternate between black and white. The triangles in the outer segment rotate their colours clockwise between grey-black-white.

Question 245: A

Each quadrant of the frame is rotated 90° clockwise.

Question 246: D

The black shapes rotate clockwise within the bottom-left quadrant. The white shapes flip vertically.

Question 247: C

Each frame has an increasing number of intersections. All shapes that touch another shape are coloured grey.

Question 248: C
As the sequence progresses, every large shape is converted into a small shape and one new large shape is added to the frame.

Question 249: D
The large square cycles between the colours grey-white-black. The large triangle cycles between black-white-grey. The smaller shapes in the top-left quadrant rotate clockwise whilst the smaller shapes in the bottom-right rotate anti-clockwise.

Question 250: B
The total number of sides for black shapes increases by one each frame.

Question 251: A
The left half of the image is flipped vertically whilst the right half is flipped horizontally. Any shape that is found in the same area in the same orientation is coloured black.

Question 252: D
Shapes with right-angles are rotated by 90° clockwise whilst those without right-angles are coloured black.

Question 253: D
White shapes increase their number of edges by one black shapes are flipped horizontally.

Question 254: D
All triangles are removed. The other shapes move to a random position

Question 255: C
The white shapes gain a larger black shape.

Question 256: B
If an arrow points at a shape it turns black, if the arrow points away from the shape then the arrow itself turns black. The orientation / organisation of shapes is not significant.

Question 257: C
White shapes rotated 90° anti-clockwise. Grey shapes are rotated 180°. Black shapes are rotated 90° clockwise.

Question 258: C
Shapes that have an identical pair (regardless of rotation) are coloured black.

Question 259: A
The entire image is rotated 90° clock-wise and then flipped vertically.

Question 260: B
The left half of the image is flipped horizontally whilst the right half is flipped vertically.

Question 261: B
The entire image is rotated 180°.

Question 262: C
Circles become triangles, triangles become squares and squares become circles. All shapes keep the same size and position.

Question 263: A
The entire image is rotated 90° anti-clockwise.

Question 264: A
Each shape is split in half along one line of reflection with one half being coloured black.

Question 265: B
Black shapes are moved upwards whilst white shapes are moved towards the left.

Question 266: A
Black shapes are moved such that they all touch each other whilst white shapes are moved such that they do not touch.

Question 267: C
The number of edges of the large black shape increases by one. The rest of the shapes are white and each have the same number of sides. The total number of shapes is equal to this same number.

Question 268: A
The stars that are pointing upwards are coloured black whilst the others are kept white.

Question 269: A
Only the shapes touching another shape remain.

Question 270: D
The colours rotate clockwise whilst the number of sides for each shape increases by one.

Question 271: A
Each individual quadrant is rotated 90° clockwise.

Question 272: C
Each quadrant is rotated 180° and the colours grey and white are switched.

Question 273: C
Every black shape is rotated 90° clockwise and coloured white whilst every white shape becomes black.

Question 274: C
Black shapes move away from each other whilst white shapes get bigger.

Question 275: D
For every shape with the same number of sides, the total decreases by one.

Question 276: D
The total number of white shapes is equal to the number of sides of the large black shape.

Question 277: A
Each shape is split into two shapes of the same colour with the same total number of edges.

Question 278: B
Every triangle is split in half through its right-angle. Grey triangles are placed by two smaller white and black triangles. The frame is rotated 180°

Question 279: C
Shapes with right-angles get larger. Shapes without right-angles get smaller.

Question 280: B
The shape in each quadrant has twice as many sides as that of the original shape.

Question 281: B
The shape is flipped diagonally and rotated 90°

Question 282: A
The entire image is rotated 90° to the right and then the new bottom half of the image is flipped vertically

Question 283: A

Each quadrant is rotated 180°.

Question 284: D

Increase the number of sides of black shapes by one and decrease the number of sides of white shapes by one.

Question 285: B

The left half of the image swaps between the shapes. The right half is rotated 180° and the colour of the shaped inverted.

Question 286: C

For every shape, there will be another shape that is not necessarily the same size or rotation.

Question 287: C

The total number of shapes within each quadrant are identical

Question 288: A

Same number of white, grey and black shapes.

Question 289: C

All objects move quadrant in a clockwise direction. Objects within squares also rotate 90° clockwise.

Question 290: C

The total number of sides is the same

Question 291: D

Every black shape becomes two different black shapes.

Question 292: A

The entire image is flipped vertically but the colours remain the same.

Question 293: A

All shapes with an odd number of sides are removed.

Question 294: C

For arrows pointing towards each other, one arrow will move next to the other. For arrows pointing away from each other, one arrow will move away from the other.

Question 295: B

The top-left and bottom-right shapes swap whilst the top-right and bottom-left shapes swap. The location of the colours remains the same

Question 296: D

The image is flipped in the diagonal axis (top-left to bottom-right)

Question 297: A

The shape in the top-left is rotated 90° clockwise; top-right 180°; bottom-right 90° anti-clockwise.

Question 298: C

The number of black shapes to calculate the number of grey shapes displayed divides the number of white shapes.

Question 299: A

Each shape is split into two shapes with half the number of sides as the original shape.

Question 300: A

Shapes with right-angles are rotated. Those without right-angles are replaced by other shapes.

Situational Judgement Answers

Q	A	Q	A	Q	A	Q	A	Q	A	Q	A
1	D	51	A	101	C	151	C	201	D	251	A
2	B	52	D	102	B	152	B	202	D	252	B
3	D	53	C	103	A	153	D	203	D	253	A
4	A	54	A	104	D	154	A	204	A	254	C
5	B	55	C	105	A	155	A	205	A	255	B
6	A	56	A	106	C	156	A	206	D	256	B
7	B	57	C	107	B	157	C	207	B	257	D
8	D	58	D	108	B	158	D	208	D	258	A
9	C	59	A	109	C	159	A	209	A	259	D
10	D	60	B	110	A	160	B	210	B	260	C
11	D	61	D	111	A	161	A	211	C	261	D
12	A	62	C	112	D	162	C	212	D	262	B
13	D	63	A	113	C	163	B	213	A	263	A
14	C	64	D	114	A	164	A	214	A	264	A
15	B	65	A	115	A	165	C	215	D	265	D
16	A	66	D	116	B	166	A	216	A	266	D
17	D	67	A	117	D	167	A	217	D	267	D
18	D	68	C	118	A	168	D	218	D	268	A
19	C	69	A	119	C	169	C	219	C	269	A
20	C	70	A	120	D	170	A	220	A	270	B
21	D	71	A	121	D	171	A	221	A	271	A
22	B	72	B	122	A	172	A	222	B	272	D
23	A	73	D	123	A	173	B	223	C	273	D
24	D	74	D	124	B	174	C	224	A	274	A
25	A	75	D	125	A	175	C	225	D	275	A
26	D	76	A	126	D	176	A	226	B	276	B
27	A	77	D	127	B	177	D	227	A	277	A
28	A	78	C	128	B	178	B	228	A	278	A
29	D	79	A	129	C	179	D	229	D	279	D
30	A	80	B	130	A	180	C	230	B	280	C
31	A	81	D	131	A	181	A	231	D	281	D
32	D	82	A	132	D	182	A	232	A	282	A
33	B	83	A	133	C	183	C	233	D	283	C
34	D	84	C	134	C	184	A	234	A	284	B
35	B	85	A	135	C	185	D	235	D	285	D
36	C	86	A	136	B	186	C	236	C	286	B
37	A	87	D	137	D	187	A	237	A	287	A
38	B	88	C	138	A	188	A	238	D	288	D
39	B	89	D	139	A	189	C	239	C	289	C
40	C	90	C	140	A	190	D	240	D	290	A
41	A	91	A	141	A	191	A	241	B	291	D
42	D	92	C	142	B	192	C	242	D	292	A
43	B	93	A	143	D	193	A	243	A	293	B
44	A	94	D	144	A	194	D	244	A	294	D
45	B	95	C	145	A	195	B	245	C	295	B
46	D	96	D	146	C	196	D	246	D	296	B
47	A	97	B	147	B	197	A	247	A	297	C
48	B	98	A	148	B	198	B	248	D	298	A
49	D	99	B	149	D	199	A	249	B	299	B
50	C	100	D	150	A	200	A	250	A	300	D

	A	A very appropriate thing to do
How appropriate is…? Scenarios: 1-21 and 41-60	B	Appropriate, but not ideal
	C	Inappropriate, but not awful
	D	A very inappropriate thing to do
	A	Very important
How important is…? Scenarios: 22-40	B	Important
	C	Of minor importance
	D	Not important at all

Scenario 1:

1. **Very inappropriate** because Jacob didn't know the entire story and this could be resolved by having a simple conversation instead. There is no need at his stage to get more people involved, especially as the doctor's behaviour has not directly affected the patient's safety or treatment.
2. **Appropriate but not ideal**, because the medical student would not be telling the doctor anything specific.
3. **Very inappropriate** because as a medical student you are a member of the health care team as well, so if there is something that is affecting the rest of the staff and patients, then it should not be ignored.
4. **Very appropriate** because the supervisor would be able to advise the student as to what they should do.
5. **Appropriate but not ideal**, whilst it makes Dr Herbert aware of the issue, it is quite confrontational and Dr Herbert may become defensive.

Scenario 2:

6. **Very appropriate**, because Dr Walker could have gotten into a bad habit and may be unaware that he hasn't been washing his hands.
7. **Appropriate but not ideal**, because Dr Walker may not pick up on the hint, although it might save George some awkwardness in having to ask Dr Walker directly.
8. **Very inappropriate** because hospitals function as a team. If George is aware of something that could potentially cause patients harm, he must try to solve the issue.
9. **Inappropriate but not awful**, because it is not addressing the situation and could make it a bigger problem than it actually is. In general, problems with doctors should be escalated to more senior doctors; problems with nurses should be escalated to more senior nurses.
10. **Very inappropriate** because Dr Walker would not have been informed and the fact that George would have witnessed it without trying to correct the problem could back fire onto him and get him into trouble if any harm were to arise.

Scenario 3:

11. **Very inappropriate** because the medical school museum would have to account for the missing bones. The bones are very valuable, and even the remains of bones could be useful as they would make them into slides or as cut sections.
12. **Very appropriate** because it acknowledges the respect to both the bones and the museum. The bones came from a real human, so cannot be treated as though they are any old piece of waste.
13. **Very inappropriate** because this would prompt an investigation and would waste a lot of money from the medical school. It would also mean that future classes may be banned from performing such projects, hindering their educational experience.
14. **Inappropriate but not awful**, because despite the fact that the curator would be informed of what occurred, it could get your colleague into trouble that could have been avoided.
15. **Appropriate but not ideal**, because the supervisor can give advice as to what to do, but it does not directly address the problem.

Scenario 4:

16. **Very appropriate** because his tutor can give him proper advice and will also be aware of any reasons behind a potentially disappointing exam mark.

17. **Very inappropriate** because it will increase tensions and result in a more stressful environment, which would hinder his progress even more.

18. **Very inappropriate** because he will end up feeling very isolated and lonely and anxious, which will also ruin both his friendships and his work progress.

19. **Inappropriate but not awful,** because he will lose out on friendships as well as go through the difficulty of finding another place to live. This could end up as a lonely option with no support network.

20. **Appropriate but not ideal** because he may find that everyone is having similar problems. It doesn't directly address the problems but Henry might find it helpful to discuss the situation with someone else.

Scenario 5:

21. **Very inappropriate** because Mark was only asked because he was mistaken for a doctor. Therefore exchanging one student for another would be an inappropriate action. Also, the other student wasn't asked.

22. **Appropriate but not ideal,** because Mark could document the results as a student and write exactly who he was. Students are allowed to perform tests, just not to administer medication.

23. **Very appropriate** because it will alert the doctor as to his mistake, and the student can be advised appropriately.

24. **Very inappropriate** because the test needs to be done, and the doctor would assume that the test had been done. Therefore the patient could be left waiting for a long time.

25. **Very appropriate** because they might be better qualified to do the test.

Scenario 6:

26. **Very inappropriate** because it is dishonest and if the truth were to emerge, she could be expelled from the medical school for such an act.

27. **Very appropriate** because they can support her and help her organise the rest of her revision.

28. **Very appropriate** because it can reassure her and give her more confidence as most people would probably feel similar.

29. **Very inappropriate** because this is also dishonest and therefore, unprofessional.

30. **Very appropriate** because she might just pass her exams and surprise herself.

Scenario 7:

31. **Very appropriate** because Dijam should not be in the hospital and is not employed to be in the hospital so can go home to recuperate.

32. **Very inappropriate** because Dijam still has contact with the patients and it is against the rules to be hungover or have alcohol in your system as a student on the wards. Dijam could get asked to do certain tasks and must make sure that he is mentally and physically competent.

33. **Appropriate** because they might send Dijam home as well, or could allow him to stay but restrict what he was allowed to do for the day.

34. **Very inappropriate** because this is a very serious violation of professionalism and joking about it is not addressing the problem.

35. **Appropriate but not ideal** because Dijam could get into more trouble than he probably ought to. That doctor also may not have witnessed Dijam in his state anyway, so it is better to inform the doctors that Dijam is shadowing than the one that is in charge.

Scenario 8:

36. **Inappropriate but not awful** as the work should be a joint effort from everyone but it won't be confrontational.

37. **Very appropriate** because Patrick's personal tutor can advise him accordingly.

38. **Appropriate but not ideal** - it will allow them to speak up but the conversation wouldn't involve Jina.

39. **Appropriate but not ideal** because the group members may not pick up on the hinting.

40. **Inappropriate** because the confrontational approach could offend Jina and not achieve what Patrick was hoping for.

Scenario 9:

41. **Very appropriate** because Joshua should never be allowed to breach confidentiality and take the patient's notes away from the hospital.

42. **Very inappropriate** because Nazia has a duty to not allow any serious breaches of confidentiality if she is aware of them.

43. **Appropriate but not ideal** because although people have been made aware, the notes have already left the hospital.

44. **Very appropriate** because your clinical advisor can give you the best advice when it is unclear what is best to do.

45. **Appropriate but not ideal** because it prevents Joshua from taking the notes away from the hospital but will cost Megan her time.

Scenario 10:

46. **Very inappropriate** as this is a clear lie and therefore grossly unprofessional.

47. **Very appropriate** because Dr Kelly cannot really explain much to Mr. Marshall without breaking the bad news to him.

48. **Appropriate but not ideal** because this may damage their relationship.

49. **Very inappropriate** because this may create confusion and being unnecessarily cryptic.

50. **Inappropriate but not awful** because although it will address his concerns, Dr Kelly is not the best person to break the news.

Scenario 11:

51. **Very appropriate** because Julia is unable to advise Mary.

52. **Very inappropriate** because Julia isn't experienced enough to assess when Mary will be able to go home.

53. **Inappropriate but not awful** because whilst technically true, it is not Julia's responsibility to advise Mary on anything without the permission of her doctor.

54. **Very appropriate** as she would then be giving correct information and also addressing Mary's concerns.

55. **Inappropriate** because Julia is not qualified and cannot answer Mary's questions.

Scenario 12:

56. **Very appropriate** because it shows that Daniel is trying his best to learn the skill with minimal fuss.

57. **Inappropriate but not awful** because Hannah needs to know that she missed someone out, but it won't solve the immediate problem.

58. **Very inappropriate** because it could have been a genuine mistake on Hannah's part, and will not make the current situation better.

59. **Very appropriate** because it might have been an issue that affected lots of people, instead of just Daniel.

60. **Appropriate but not ideal**. This shows that Daniel cares about his assessment but also means extra hassle for the doctor.

Scenario 13:

61. **Very inappropriate** because her friend will have spent a lot of money on the flights and will be left to fend for herself or not be able to go.

62. **Inappropriate but not awful** because the holiday will not be enjoyable for Helen's friend, and Helen will also not benefit from having a fun experience.

63. **Very appropriate** because Helen also needs a break and being able to go for part of the holiday will be rewarding without having to miss out completely, and will leave enough time to prepare for the exams too.

64. **Very inappropriate** because Helen will be jeopardizing her chances of remaining in medical school.

65. **Very appropriate** because she will not be disappointing her friend as much, and will still be able to do her revision.

Scenario 14:

66. **Very inappropriate** because Celia's parents have not chosen to isolate her on purpose.
67. **Very appropriate** because Celia can meet new people without having to worry about insulting her parents.
68. **Inappropriate but not awful** as although she will be closer to her friends, she may turn into an unwelcome guest quite quickly.
69. **Very inappropriate** because this will affect her mental health and her relationships, and will probably lead to underperformance in her studies as well
70. **Very appropriate** because it shows that she understands her family's situation whilst attempting to address her own predicaments.

Scenario 15:

71. **Very appropriate** – Xun should try to do it to the best of his ability in the remaining time.
72. **Appropriate but not ideal** – it is unlikely that the deadline will be unchanged but there is always space for honesty and he might just be lucky.
73. **Very inappropriate** because this action will affect his final grade.
74. **Very inappropriate** because this is plagiarism and both Xun and his friend would be penalised for it.
75. Very inappropriate because he would be lying to the assessment office and could get into a lot of trouble for it.

Scenario 16:

76. **Very appropriate** because he will be showing that he respects the hospital, infection control, and the patients.
77. **Very inappropriate** because this will set a bad first impression to his patients and colleagues – made worse by his refusal.
78. **Inappropriate but not awful.** Nahor may set a bad first impression but there is a small chance that he might be allowed to keep it.
79. **Very appropriate** because he can still enjoy keeping his hair in an individual style without looking unprofessional.
80. **Appropriate but not ideal.** Whilst this is a proactive approach, it is such a minor issue that it is not worth wasting the dean's time about. In addition, the answer is very likely to be 'no'.

Scenario 17:

81. **Very inappropriate** because Dr Patel will not receive the email until he has handed over to the doctor that they are waiting for and it could affect patient safety if Charles leaves.
82. **Very appropriate** because Charles knew he was going to be on call. If the doctor was running late then it would be different, but as this was planned Charles should wait for his commitment to finish first.
83. **Very appropriate** because it addresses the problem directly and Dr Patel would then know that Charles was leaving.
84. **Inappropriate but not awful** because Charles should not rely on messengers as they aren't as reliable as telling Dr Patel himself.
85. **Very appropriate** because the doctor that is taking over will be aware of the situation and can brief Dr Patel appropriately.

Scenario 18:

86. **Very appropriate** because Archie is being truthful and professional without ignoring the patient and still addresses her concerns.
87. **Very inappropriate** because Archie is not qualified to answer questions regarding management from a patient.
88. **Inappropriate but not awful** because whilst Archie is not giving any harmful information directly, the patient is unlikely to find his advice useful and it doesn't really address the problem.
89. **Very inappropriate** because Archie doesn't have the expertise necessary to give medical advice – with or without the use of Google.
90. **Inappropriate but not awful** because it is doesn't address the problem directly but at least isn't going to make the situation worse.

Scenario 19:

91. **Very appropriate** because they can advise Matthias on the best possible care.
92. **Inappropriate but not awful** because it may slow his recovery and may be dangerous. However, he would at least know if he was able to cope with the activity.
93. **Very appropriate** because it shows that he is being proactive and does not want to waste time.
94. **Very inappropriate** because the medical school might assume that he is being lazy and not going to hospital as he isn't committed.
95. **Inappropriate but not awful** because the medical school should find out from Matthias- not his friends.

Scenario 20:

96. **Very inappropriate** because Jessie may be the only person who notices this and is therefore in a position to address it. If nothing is done, Gemma may become very isolated.
97. **Appropriate** but not ideal because Gemma will probably not want to talk about her eating disorder and avoid the topic in the future.
98. **Very appropriate** because it shows that Jessie can be trusted and is there for Gemma, without being confrontational.
99. **Appropriate but not ideal** as whilst it may make Gemma's parents aware – it is likely to strain their friendship.
100. **Very appropriate** because they will have seen this scenario before, and will know how to respond to it.

Scenario 21:

101. **Inappropriate, but no awful** – although this will allow her to revise for her exams, her friend will have spent a lot of money on the flights and will be left to fend for herself or not be able to go.
102. **Appropriate but not ideal-** Helen would be addressing both issues but given the circumstances, her revision may suffer or her friend may not enjoy the holiday as much.
103. **Very appropriate** because Helen also needs a break and being able to go for part of the holiday would be rewarding and shouldn't affect her revision as long as she comes back early enough.
104. **Very inappropriate** because Helen will be jeopardizing her chances of remaining in medical school.
105. **Very appropriate** because she will not be disappointing her friend as much, and will still be able to do her revision.

Scenario 22:

106. **Of minor importance** because the task should be joint effort.
107. **Important,** because this means that their grade is significant, and Sean will want to do as well as possible.
108. **Important** because they should learn how to work together to prevent future problems with their group work.
109. **Of minor importance-** although this might be troubling Daniel, he shouldn't let his social life affect his work life.
110. **Very important** because Daniel maybe used to not pulling his weight, and will have to be informed that he needs to contribute more.

Scenario 23:

111. **Very important,** because if the essay counts towards Tanya's final mark for the year then she would want to do very well.
112. **Not important at all** – different people learn in different ways and every clinic is different.
113. Of minor importance, because Dr Garg will be assessing her at the end of the term.
114. **Very important,** because if this opportunity is available at another time, then missing this particular clinic is not particularly disastrous for Tanya's learning.
115. **Very important,** because Tanya can complete the essay in time for the clinic, then she would not be compromising her learning.

Scenario 24:

116. **Important**, because the doctors that will be assessing Caroline will associate her with the dishevelled looking first year students and it may look unprofessional.

117. **Not important at all**, because they will still be representing the medical school and as such have a duty towards acting properly.

118. **Very important,** because the first year students are therefore expected to dress appropriately and professionally.

119. **Of minor importance,** because if she needs to ask them to dress more appropriately, they shouldn't get offended.

120. **Not important at all** – the students are representing the medical profession and must appear appropriately presentable regardless of when their examinations are.

Scenario 25:

121. **Not important at all** – timetabled clinical commitments should always come first (without extenuating circumstances).

122. **Very important**, because he may have to show a sports contribution to the medical school if he is on a scholarship. If this is the case, his firm head might understand, and rearrange the teaching sessions.

123. **Very important,** because he may not be allowed to finish his degree if he chooses to miss the sessions and not inform the consultant.

124. **Important** because this might allow him to rearrange the teaching sessions.

125. **Very important** because if he is missing invaluable teaching then this will affect his training and disadvantage him in the long run.

Scenario 26:

126. **Not important at all** – patient satisfaction in important and challenging bad practice does not reflect badly on students.

127. **Important** because the patient might be reluctant to ask questions or comply with the treatment if they are uncomfortable.

128. **Important** because if it is not a regular thing, then perhaps it is because Dr Davison is particularly stressed one day and might not be aware that he is swearing.

129. **Of minor importance** because whether or not they are aware of his swearing, it still makes the patient feel uncomfortable.

130. **Very important** because that means that the patient was either comfortable with the doctor previously, or that they were uncomfortable from before.

Scenario 27:

131. **Very important** because the patient could be subjected to avoidable infections, which could be life threatening.

132. **Not important at all** because the risk of the patient receiving an infection is worth the time required for the theatre staff to get more equipment ready.

133. **Of minor importance,** because he is only a medical student so can be forgiven for making mistakes. The repercussions of not saying anything could be far more serious.

134. **Of minor importance** because there is still a risk of infection if he is not sterile but the trolley is. He could still transmit infections.

135. **Of minor importance** because it is still important to protect the patient first and foremost.

Scenario 28:

136. **Important** because if the patient really is in pain then action should be taken. However, given that the patient is addicted to pain medication, it becomes less important.

137. **Not important at all** because her demands for pain medication are irrelevant to the staff, especially since she has just been reviewed by the doctors and nurses.

138. **Very important** because Freddie cannot administer any medications.

139. **Very important** because this means that the pain medication is not working or she is asking for it unnecessarily.

140. **Very important** because the patient has been reviewed and is therefore unlikely to need anything else so soon.

Scenario 29:

141. **Very important** because this is the key issue and the most important one for patient outcomes.

142. **Important** because letting Claire know this would put Claire at ease but it is not an essential piece of information to convey.

143. **Not important at all** as it is irrelevant to Claire's operation.

144. **Very important,** because her anxiety could affect her recovery if she is not completely comfortable with the surgeon.

145. **Very important** because if this is true, it will help Claire's recovery.

Scenario 30:

146. **Of minor importance** because Annabel should respond according to the situation, not because of who is involved. However, she should be wary that her actions are not in revenge.

147. **Important** because it will show that he usually works hard, whether or not he receives the help.

148. **Important** because the help from his rugby friends would not count as cheating.

149. **Not important at all** – Annabel might be jealous that she doesn't have this support network, but that should not influence her actions.

150. **Very important** because it severely disadvantages students who don't have access to the answers.

Scenario 31:

151. **Of minor importance** – whilst it won't stop him progressing with his degree, professionalism is a very important thing, and the medical school might want to investigate it further.

152. **Important,** because Matthew would have evidence of the public transport being delayed.

153. **Not important at all,** because the punctuality of the rest of his class does not reflect on Matthew at all.

154. **Very important,** because if Matthew can prove that his punctuality is usually satisfactory, then he can be more readily forgiven or excused for being late on the one occasion.

155. **Very important,** because Matthew should maintain a good rapport with his teacher if they are to see each other regularly for the next year.

Scenario 32:

156. **Very important,** because the week will be a vital part of her learning experience, and she will miss out if she misses the opportunity.

157. **Of minor importance** because whilst it may not cause seriously harm healthy people, it could seriously harm ill patients.

158. **Not important at all** because the issue here is of infection control – not if Michaela can function in the hospital.

159. **Very important** because if a patient became ill, Michaela would be unable to say that she had not been instructed appropriately.

160. **Important,** as it would be a useful opportunity but not very important because the information is just her friend's opinion.

Scenario 33:

161. **Very important,** as this may be Jenny's only attempt to see what the exams are going to be like.
162. **Of minor importance** – Jenny's education is more important than losing out on money for a ski trip. However, it is still important, especially depending on what her financial situation is, but is not as relevant when deciding on how important the exams are.
163. **Important,** as Jenny would not have to worry about missing them or swapping her exam dates.
164. **Very important,** because if the university agrees to swap their exam dates then the issue would be resolved.
165. **Of minor importance,** although she has done well on exams so far, there is no guarantee that it would be the same for this set of mocks. It is still something to bear in mind whilst making her choice.

Scenario 34:

166. **Very important,** because it shows that he was very interested and proactive in the subject. It can also be used as evidence that Luke should be reconsidered for the project.
167. **Very important** because Architha may be willing to help Luke in his appeal as she could also gain from it.
168. **Not important at all,** because Luke should try to achieve the best grade possible with whichever project he was given.
169. **Of minor importance,** although Luke will find it difficult to concentrate on the project that he was allocated, personal preferences should not hinder his overall performance.
170. **Very important,** because the medical school can show that took his preferences into account.

Scenario 35:

171. **Very important,** because the mark will reflect on her ability for the rest of her life.
172. **Very important,** because this act might prevent Lucinda from getting the job she wants later on in her medical career.
173. **Important,** because it is a short period of time and if their relationship ends then Lucinda would have waster a year.
174. **Of minor importance-** although she is used to being with people that do well, they shouldn't affect her decision.
175. **Of minor importance,** although they are in a relationship, she needs to make the decision herself rather than be pressured into it.

Scenario 36:

176. **Very important,** because Shiv will save a lot of money if he skips the last day.
177. **Not important at all,** because the content of the assessment is not the issue but rather its timing.
178. **Important,** because this shows that Shiv is a good student. It also means that the doctor may have already formed an opinion for Shiv, and might be more likely to be lenient if he misses the final assessment.
179. **Not important at all,** because it will have no impact on Shiv's grade.
180. **Of minor importance** – although Shiv would like to meet with his girlfriend, he will be able to see her eventually so this isn't important in deciding **when** to leave.

Scenario 37:

181. **Very important** because if Jazzmynne won't get another chance to go on tour then it may be worth considering if she can miss classes.
182. **Very important,** because her learning is also important, and if she misses this rotation then she may not necessarily get the chance to catch up again.
183. **Of minor importance,** because Jazzmynne is an adult and if she feels as though she can handle the workload then she should be able to make her decisions herself. However, she should still keep her parent's concerns in mind.
184. **Very important,** because if Jazzmynne sometimes struggles to keep up with her workload without having any help, then with the extra stress of missing lots of work, she might struggle a lot.
185. **Not important at all** as Jazzmynne should not make the decision based on how much she would gain from the tour not because of what her friends decide.

Scenario 38:

186. **Of minor importance.** The changes may not come into action for Ellen to see but if she still feels strongly about the paper then it shouldn't change her actions.
187. **Very important**, as this shows that the university will take the opinions of the students into account.
188. **Very important,** because if the general stories are changed completely, then the majority of student readers may stop reading the papers. This could impact the newspaper's finances adversely.
189. **Of minor importance,** because the principal's stories can still be incorporated into the newspapers without abolishing the papers on the students' social lives.
190. **Not important at all** as this isn't a personal issue and should be handled professionally instead.

Scenario 39:

191. **Very important,** because the student bars should be safe, controlled spaces for the students to enjoy themselves.
192. **Of minor importance** – whilst extra money would be useful, the decision to appeal should be based on the needs of the student body rather than serving a selfish agenda.
193. **Very important,** because this is another factor that could bring their satisfaction ratings down further.
194. **Not important at all** – the issue is not about the cost but about the waiting times.
195. **Important,** because expanding the bar could bring in more money for the university.

Scenario 40:

196. **Not important at all,** because this should be handled amiably rather than making it personal.
197. **Very important,** because Phil does not have a chance of running again, but Olivia does.
198. **Important,** because it shows that Olivia is dedicated to the student union. It is not very important as there will be other factors that will differentiate Olivia and Phil.
199. **Very important,** because this excludes any chance of compromise with position allocation.
200. **Very important,** because it shows that Olivia has already compromised with the position previously, and will be unlikely to want to compromise again

Scenario 41:

201. **Very inappropriate-**This shows a lack of honesty and integrity and it is unfair to other students.
202. **Very inappropriate-** Although this appears to be a fair process as all the students are receiving the exam, it undermines the exam process and displays a lack of integrity in practice.
203. **Very inappropriate-** As a medical student you should take initiative and escalate situations appropriately. When you see that there is something that can affect the rest of the students and the medical school, you should involve those at a senior level
204. **Very appropriate-** The supervisor would be able to advise the student on what to do and would be able to help them in escalating the issue if required.
205. **Very appropriate-** The medical school will have a protocol for reporting situations such as these, and can deal with the situation accordingly.

Scenario 42:

206. **Very inappropriate-** Kate does not know the circumstances surrounding the missing money. It would be inappropriate to involve the police without the patient's permission and without exploring other avenues such as asking the senior nurses or look for the money herself.
207. **Appropriate but not ideal-** Kate needs to be careful not to sound patronising, as the patient will know that the wards are very exposed. However, it is good she offers them reassurance.
208. **Very inappropriate-** This is very patronising to the patient. Kate is accusing them of lying without understanding the full story. Also she has not provided any re-assurance that she will follow this up with more senior members of the department.
209. **Very appropriate-** Kate has taken some further details, she has re-assured the patient and has escalated the issue to the right person. Also the nurse in charge will be aware of the hospital processes and would be able to provide the patient with more information and assurance.

210. **Appropriate but not ideal-** Alerting other students of the theft is appropriate so they can keep their possessions safe and be on the lookout for the money, however it does not deal with the patient's alleged theft.

Scenario 43:

211. **Inappropriate but not awful-** Sam should not accuse Andrew of possession before he clarifies the situation. However, escalating the problem to his supervisor is appropriate.

212. **Very inappropriate-** Ignoring the situation, especially when Andrew may need help and support is inappropriate. There are also future implications to consider. Agreeing with Andrew to get rid of the drugs may lead Andrew to feel that taking drugs is ok.

213. **Very appropriate-** The supervisor would be able to provide advice to Sam on what he should do next.

214. **Very appropriate-** Offering support to Andrew would provide reassurance and a possibility to confide in a friend. Recommending professional help means that Sam recognises his own limitations and has signposted Andrew down the correct pathway.

215. **Very inappropriate-** Doing nothing is totally inappropriate. Seeing patients under the influence of drugs may put patient safety at risk and totally undermines a doctor's position of trust.

Scenario 44:

216. **Very appropriate:** Anna has recognised her limitations, informing someone who is senior who has more experience will ensure that the situation is dealt with more promptly and effectively. Patient safety is also maintained, and will ensure that the junior doctor receives professional help if he needs it.

217. **Very inappropriate-**Anna should not ignore the situation, the doctor is obviously drunk and he may put patients in danger. As a medical student if she sees there is something wrong in medical practice she needs to act on it.

218. **Very inappropriate-** Anna needs to recognise the limitations of her expertise and allowing the doctor to carry out his duties under the influence of alcohol can place patient safety at risk.

219. **Inappropriate but not awful-** It is good to reflect on difficult situations in practice, however this has not provided a solution to the current situation.

220. **Very appropriate -** Anna has recognised that the doctor is drunk and won't be able to carry out his duties. Advising him to inform the team so they are aware of the situation is very appropriate. More senior members of the team will be able to follow this up and ensure that the ward is appropriately staffed. .

Scenario 45

221. **Very appropriate-** Escalating the problem to the supervisor will enable the supervisor to provide solutions to the problem, support the group and encourage Sarah.

222. **Appropriate but not ideal-** Seeing how other group members feel about the current situation is appropriate, however it may lead to spreading gossip about Sarah.

223. **Inappropriate but not awful-** Informing Sarah that he has noticed her lateness and lack of contribution is an appropriate course of action. However the approach used is very patronising, furthermore there may be an underlying reason for her lateness that Sean should have asked about.

224. **Very appropriate-** Clarifying the situation with Sarah and making her aware that her lateness and contribution have not gone unnoticed allows her to increase her contribution and change her behaviour. Providing support may mean that he is able to remedy the current situation.

225. **Very inappropriate-** This is unfair to the other group members, in addition Sarah may be going through personal problems that she needs support for.

Scenario 46

226. **Appropriate but not ideal-** The nurse may not be aware of whom Mary is and Mary may come across as patronising, so the nurse may not be inclined to listen to her.

227. **Very appropriate-** Reporting the situation to a senior member is appropriate. The best person to report it to is the nurses immediate senior.

228. **Very appropriate-** Reporting the situation to the consultant is appropriate as they would be able to provide Mary with advice on what to do, or take it to the nurse's immediate senior themselves.

229. **Very inappropriate-** The nurse has acted unprofessionally, taking medications from the ward may put patients at risk, therefore the nurse needs to be made aware of the effects of her actions, and her need to uphold professional values.

230. **Appropriate but not ideal-** Reflecting on situations at medical placements is integral, as it will allow Mary to become a better future doctor. However, she has not dealt with the current situation of the nurse taking the medication.

Scenario 47

231. **Very inappropriate-** This is a matter between the consultant and Alan. Involving the nurse is inappropriate as she may inform the consultant, but this would involve an unnecessary party and may erode trust between the consultant and Alan.

232. **Very appropriate-**Alan's issue is with the consultant, so directly discussing it with him may allow Alan to gain a good relationship with the consultant and allow them to come to a joint decision regarding the remaining sessions.

233. **Very inappropriate-** This will only inflame the situation. The consultant may not realise that he is being harsh on Alan and therefore openly arguing in front of the patient is unprofessional and only sour their relationship.

234. **Very appropriate-** Alan is involving the relevant authority. It is his supervisor who is responsible for his educational needs, and his supervisor is in a position to provide advice at an appropriate level.

235. **Very inappropriate-** In this case, Alan is involving a patient in a conflict which is really none of their concern Alan is also undermining the Consultant's credibility behind their back.

Scenario 48

236. **Inappropriate but not awful-** The nurse has specifically asked Harry if he could have a word with his fellow student, because she may have felt that it is a situation that could be solved at the student level without involving doctors. Harry may feel that this is not his responsibility.

237. **Very appropriate-** Addressing the issue with the colleague in question avoids public humiliation and a solution to the problem may be reached without involving someone else such as the consultant.

238. **Very inappropriate-** Harry is being dishonest to the nurse especially when he has promised to speak to Joe.

239. **Inappropriate but not awful-** Harry is not publically humiliating Joe and he has not ignored the issue, but sending him an anonymous note is very cowardly and may upset Joe.

240. **Very inappropriate-**This action will achieve nothing other than publicly humiliating Joe in front of other students.

Scenario 49

241. **Appropriate but not ideal-** It is appropriate to be honest to the consultant and to refuse to add the names unless they have contributed to the report.

242. **Very inappropriate-** This is dishonest and unprofessional, it doesn't matter what type of publication; it is a matter of personal ethics that Helen has to her patients and society.

243. **Very appropriate-**Escalating the problem to the medical student supervisor is appropriate as he would be able to support and provide Helen with the advice needed.

244. **Very appropriate-** Helen could clarify if the added names have actually contributed to the report and to discuss her concerns.

245. **Inappropriate but not awful-** This does not solve the situation; not publishing avoids being an accessory to fraudulent activity, however Helen is losing a chance to add an achievement to her own CV.

Scenario 50

246. **Very inappropriate-**John is being presumptive here, the books may be Clare's own, and Clare may have borrowed the books from the library and forgot to return them.

247. **Very appropriate-** John has clarified the situation with Clare and has asked her to return the books, so has provided a solution. Medical school library books are a resource for all students so now other students would be able to use the missing books.

248. **Very inappropriate-** There is no need to contact the police in this situation, John would be wasting their time and resources.

249. **Appropriate but not ideal-** The medical school library would now know that they do not need to purchase more copies of the missing books, and they may follow it up independently, but John has not ensured that Clare would return them.

250. **Very appropriate-** If John is not sure of how to tackle the situation it would be best to discuss it with his supervisor as he would give him the advice needed.

Scenario 51

251. **Very appropriate-** Speaking to Anne is the best approach to take as Anne may not be aware of how Nadia is feeling. Anne and Nadia may also be able to find an arrangement of providing teaching sessions when the ward is quite.

252. **Appropriate but not ideal-**Nadia needs to take initiative for her own learning. If she feels she is not making the most of her experience then she needs to inform someone, and speaking sooner rather than later would obviously be better for her, as the situation may not improve on its own. However she may feel that it is too early and the situation may improve on its own as she is new to the ward.

253. **Very appropriate-** Her supervisor would be a good person to speak to, to get advice and help

254. **Inappropriate but not awful-** Nadia needs to spend time on the wards during her medical school years, as this will give her invaluable patient experience, However, she may feel that her time is spent more efficiently in the library.

255. **Appropriate but not ideal-** Nadia has taken initiative for her learning, however Anne may be offended if she has not discussed this with her prior to finding another doctor. Also the other doctors may have other students assigned or other commitments to attend to.

Scenario 52

256. **Appropriate but not ideal-** Reprimanding Laura is patronising, however refusing to sign the book is the right thing to do. Students should not be signing each other's books as it undermines the value of the clinical experience. However, offering tips for the future is a way of being supportive.

257. **Very inappropriate-** Despite seeing Laura attending the sessions, this is dishonest. It would be considered forgery as doctors are required to sign.

258. **Very appropriate-** Directing Laura to the right person who is meant to be signing the book is appropriate as they know the appropriate steps to take.

259. **Very inappropriate-** Even if Laura has demonstrated the skill, students should not be signing assessments for one another.

260. **Inappropriate but not awful-** Reporting Laura to the medical school without clarifying the situation or trying the solve it is a bit extreme, however forgery is a great academic offence.

Scenario 53

261. **Very inappropriate-** There must be a reason why it took Steven three attempts to cannulate the patient, Emma should work within her limitations.

262. **Appropriate but not ideal-** It would be best to clarify the situation with Steven before escalating, and it would be best to escalate to his direct senior. However, alerting the nurse in charge about Steven's behaviour may avoid the situation escalating to a complaint by the patient.

263. **Very appropriate-**Steven has left Emma in an awkward position. It's best to go and find him and check everything is okay- he may have gone to get more cannulation equipment and was planning to come back.

264. **Very appropriate-** Escalating the situation to a more senior doctor is appropriate, they would be able to provide Emma with advice on what to do.

265. **Very inappropriate-** Emma is undermining Steven as a doctor, and this would affect patient-doctor trust.

Scenario 54

266. **Very inappropriate-** Jill should remain professional in her manner even if the patient has said something offensive, and retaliating could make the situation worse.

267. **Very inappropriate-** As a medical student Jill is part of the team. Ignoring the situation shows she is not a good team player. The patient may also view her silence as an excuse to continue to make these racist remarks.

268. **Very appropriate-** Racist comments are not tolerated in any workplace, so speaking to the patient politely maintains professionalism and shows that Jill has taken initiative.

269. **Very appropriate-** Discussing the matter with the consultant in charge is appropriate, he will provide Jill with the right advice and can help in de-escalating the situation.

270. **Appropriate but not ideal-** Reporting the matter straight to the registrar is appropriate as he would know the hospital protocol, especially since he is the one treating the patient. However it may be better to report it to someone else in case he takes offense.

Scenario 55

271. **Very appropriate-** Depending on how the other members of the group feel, he could then get a better understanding of the group dynamics and the appropriate steps to take.

272. **Very inappropriate-** This is being very patronising and unfair to Hannah, she needs to contribute to the group project as well.

273. **Very inappropriate-**All the group members need to contribute equally to the project. This would also be unfair to the other members of the group.

274. **Very appropriate-** The supervisor would be able to provide advice and he would usually know the appropriate steps to take in regards to involving Hannah.

275. **Very appropriate-** Speaking to Hannah directly to let her know of how you are feeling and maybe offer both support and guidance.

Scenario 56

276. **Appropriate but not ideal-** It would be best for her colleague to go home. She may become more distressed, and won't be able to do her work effectively, however she will need to notify her supervisor as she may need a few more days off.

277. **Very appropriate-** Offering support to a colleague who is going through a difficult time would be appropriate.

278. **Very appropriate-** This is very appropriate as he would be able to provide Jenny with advice and provide her colleague with support.

279. **Very inappropriate-** Jenny is not being supportive at all, this may cause her colleague to become more distressed.

280. **Inappropriate but not awful-** Jenny may have thought that leaving her colleague and giving her a few minutes will help her to calm down, but her colleague may feel alone and unsupported.

Scenario 57

281. **Very inappropriate -** This is a breach of patient confidentiality. Patient details should not be discussed in public places.

282. **Very appropriate-** Maya is within a group of friends and she should remind her friends of patient confidentiality and their duty to protect it.

283. **Inappropriate but not awful-** Maya is trying to changing the subject of the conversation, however she may not be successful and she has not highlighted to her friends that what they are doing is wrong.

284. **Appropriate but not ideal-** Maya's friends need to made aware of their actions at this stage, not saying anything means her friends will continue breaching patient confidentiality until they are called by the medical school. This would also cause unnecessary worry for her friends, and suspicion of her, especially when the matter could have been solved between friends.

285. **Very inappropriate-** By excusing herself she has taken no action to make her friends aware that they are breaching patient confidentiality, so they will continue discussing patient cases, which is not acceptable.

Scenario 58

286. **Appropriate but not ideal-** The supervisor would be able to provide Michael with advice on how to proceed, but Simon may not have wanted anyone else to know.

287. **Very appropriate-** Simon has been advised to speak to his supervisor so that the supervisor could provide him with the support he needed. He will also be aware of the medical school procedures so he would be able to assist Simon directly.

288. **Very inappropriate-** Simon has confided in Michael seeking support, mocking him may make him feel worse about himself.

289. **Inappropriate but not awful-** Simon has confided in Michael and has asked him not to tell anyone. Michael feels if more people know then they might help him, but he has broken Simon's confidence and lost his trust.

290. **Very appropriate-** Advising him to seek help from the right channels is the right thing to do. Seeking medical advice is also a duty of all students and doctors if they feel their physical or mental health has an impact on their work.

Scenario 59

291. **Very inappropriate-** The doctor is in breach of hospital regulations and his moral obligations; he needs to be made aware that what he is doing is wrong.

292. **Very appropriate-** Mark's supervisor would be aware of the guidelines for similar issues so he would be able to advise Mark on what to do and to support him.

293. **Appropriate but not ideal-** It is good to speak to the doctor directly, however, confrontation can make the situation worse, it is best to approach the doctor and clarify the situation; it may have been something innocuous such as a computer virus.

294. **Very inappropriate-** Mark has not clarified the situation with the doctor; this may anger the doctor and may lead to patients losing trust in doctors.

295. **Appropriate but not ideal-** Contacting the hospital IT services to make them aware that these website are accessible so they can firewall these sites would be good, however it does not lead to immediately solving the situation of the doctor watching pornography in a public place.

Scenario 60

296. **Appropriate but not ideal-** She is offering her apologies, however is involving another student and this is not the appropriate step to take.

297. **Inappropriate but not awful-** The change in seminar times was very short notice, so other students may be off sick however Alice would have to justify her absence at a later date.

298. **Very appropriate-** Discussing the issue with the seminar leader, the person who knows best, is proactive and shows that Alice is taking initiative and she has not involved a third party.

299. **Appropriate but not ideal-** Alice has realised that she has to attend the teaching session, it is good to have a work and social life balance so cancelling the dinner plans is not ideal.

300. **Very inappropriate-** The change in seminar times is short notice however; there are ways that Alice could approach the issue without being rude.

Final Advice

Arrive well rested, well fed and well hydrated

UKCAT is an intensive test, so make sure you're ready for it! Unfortunately you can't take water into the test, so be well hydrated before you go in. Make sure you're well rested and fed in order to be at your best!

Ask for extra whiteboards

If you're running short of whiteboard space and need another, **plan ahead and put up your hand in good time**. You don't want to be stuck with nowhere to write, waiting for someone to notice and come to help you! Act early.

Move On

If you're struggling, move on. Every question has equal weighting and there is no negative marking. **In the time it takes to answer on hard question, you could gain three times the marks** by answering the easier ones. Be smart to score points

Using your UKCAT Score

Different medical schools use UKCAT in different ways – so use this to your advantage! If you score well on UKCAT, you may help your chances of success by choosing medical schools which use it as a major component of the application process. On the other hand, if your score isn't so good, you may prefer to opt for universities that don't look at your UKCAT score, or ones that use it as a more minor consideration.

By making choices <u>after</u> finding out your score, you can increase your chance of getting a place.

Afterword

Remember that the route to a high score is your approach and practice. Don't fall into the trap that *"you can't prepare for the UKCAT"*– this COULD NOT be further from the truth. With knowledge of the test, some useful time-saving techniques and plenty of practice you can dramatically boost your score.

Work hard, be persistent and do yourself justice.

Good luck!

Acknowledgements

I would like to express my gratitude to everyone who helped make this book a reality, especially the Tutors who shared their expertise in compiling this huge collection of questions and answers. Special thanks go to my friends and family for their endless care and emotional support. Lastly, this would be complete without thanking *Rohan*, whose tireless work and good humour has made everything possible.

David Salt

About UniAdmissions

UniAdmissions is the UK's number one university admissions company, specialising in **supporting applications to Medical School and to Oxbridge**.

Every year, *UniAdmissions* helps thousands of applicants and schools across the UK. From free resources to these *Ultimate Guide Books* and from intensive courses to bespoke individual tuition, *UniAdmissions* boasts a team of **300 Expert Tutors** and a proven track record of producing great results.

We also run an **access scheme** that provides free support to students who are less able to pay. To find out more about our support like intensive **UKCAT courses** and **UKCAT tuition**, check out our website **www.uniadmissions.co.uk/ukcat**

Your Free Book

Thanks for purchasing this Ultimate Guide Book. Readers like you have the power to make or break a book – hopefully you found this one useful and informative. If you have time, *UniAdmissions* would love to hear about your experiences with this book.

As thanks for your time we'll send you another ebook from our Ultimate Guide series absolutely <u>FREE</u>!

How to Redeem Your Free Ebook in 3 Easy Steps

1) Find the book you have either on your Amazon purchase history or your email receipt to help find the book on Amazon.

2) On the product page at the Customer Reviews area, click on 'Write a customer review'

Write your review and post it! Copy the review page or take a screen shot of the review you have left.

3) Head over to **www.uniadmissions.co.uk/free-book** and select your chosen free ebook! You can choose from:

- The Ultimate UKCAT Guide – 1250 Practice Questions
- The Ultimate BMAT Guide – 600 Practice Questions
- The Ultimate TSA Guide – 300 Practice Questions
- The Ultimate LNAT Guide – 400 Practice Questions
- The Ultimate NSAA Guide – 400 Practice Questions
- The Ultimate ECAA Guide – 300 Practice Questions
- The Ultimate ENGAA Guide – 250 Practice Questions
- The Ultimate PBSAA Guide – 550 Practice Questions
- The Ultimate FPAS SJT Guide – 300 Practice Questions
- The Ultimate Oxbridge Interview Guide
- The Ultimate Medical School Interview Guide
- The Ultimate UCAS Personal Statement Guide
- The Ultimate Medical Personal Statement Guide
- The Ultimate Medical School Application Guide
- BMAT Past Paper Solutions
- TSA Past Paper Worked Solutions

Your ebook will then be emailed to you – it's as simple as that!

Alternatively, you can buy all the above titles at **www.uniadmisions.co.uk/our-books**

UKCAT Intensive Course

If you're looking to improve your UKCAT score in a short space of time, our **UKCAT intensive course** is perfect for you. It's a fully interactive seminar that guides you through all 5 sections of the UKCAT.

You are taught by our experienced UKCAT experts, who are Doctors or senior Oxbridge medical tutors who excelled in the UKCAT. The aim is to teach you powerful time-saving techniques and strategies to help you succeed for test day.

➢ Full Day intensive Course
➢ Copy of our acclaimed book "The Ultimate UKCAT Guide"
➢ Full access to extensive UKCAT online resources including:
➢ 2 complete mock papers
➢ 1200 practice questions
➢ Online on-demand lecture series
➢ Ongoing Tutor Support until Test date – never be alone again.

Timetable:

➢ **1030 - 1200:** Section 1
➢ **1200 - 1330:** Section 2
➢ **1330 - 1400:** Lunch
➢ **1400 - 1515:** Section 3

➢ **1515 - 1615:** Section 4
➢ **1615 - 1700:** Section 5
➢ **1700 - 1730:** Summary
➢ **1730 - 1800:** Questions

The course is normally £195 but you can get **£ 10 off** by using the code *"BKTEN"* at checkout.

www.uniadmissions.co.uk/ukcat-crash-course

£10 VOUCHER:

BKTEN

BMAT Intensive Course

If you're looking to improve your BMAT score in a short space of time, our **BMAT intensive course** is perfect for you. It's a fully interactive seminar that guides you through sections 1, 2 and 3 of the BMAT.

You are taught by our experienced BMAT experts, who are Doctors or senior Oxbridge medical tutors who excelled in the BMAT. The aim is to teach you powerful time-saving techniques and strategies to help you succeed for test day.

➢ Full Day intensive Course
➢ Copy of our acclaimed book "The Ultimate BMAT Guide"
➢ Full access to extensive BMAT online resources including:
➢ 4 complete mock papers
➢ 600 practice questions
➢ Fully worked solutions for all BMAT past papers since 2003
➢ Online on-demand lecture series
➢ Ongoing Tutor Support until Test date – never be alone again.

Timetable:

➢ **1000 – 1030:** Registration
➢ **1030 – 1100:** Introduction
➢ **1100 – 1300:** Section 1
➢ **1300 – 1330:** Lunch

➢ **1330 – 1600:** Section 2
➢ **1600 – 1700:** Section 3
➢ **1700 – 1730:** Summary
➢ **1730 – 1800:** Questions

The course is normally £195 but you can get **£ 10 off** by using the code *"BKTEN"* at checkout.

www.uniadmissions.co.uk/bmat-course

£10 VOUCHER:

BKTEN

Medicine Interview Course

If you've got an upcoming interview for medical school – this is the perfect course for you. You get individual attention throughout the day and are taught by Oxbridge tutors + senior doctors on how to approach the medical interview.

- ➢ Full Day intensive Course
- ➢ Guaranteed Small Groups
- ➢ 4 Hours of Small group teaching
- ➢ 2 x 30 minute individual Mock Interviews + Written Feedback
- ➢ Full MMI interview circuit with written feedback
- ➢ Ongoing Tutor Support until your interview – never be alone again

Timetable:

- ➢ **1000 - 1015:** Registration
- ➢ **1015 - 1030:** Talk: Key to interview Success
- ➢ **1030 - 1130:** Tutorial: Common Interview Questions
- ➢ **1145 - 1245:** 2 x Individual Mock Interviews
- ➢ **1245 - 1330:** Lunch
- ➢ **1330 - 1430:** Medical Ethics Workshop
- ➢ **1445 - 1545:** MMI Circuit
- ➢ **1600 - 1645:** Situational Judgement Workshop
- ➢ **1645 - 1730:** Summary and Questions

The course is normally £295 but you can get £35 off by using the code *"BRK35"* at checkout.

www.uniadmissions.co.uk/medical-school-interview-course

£35 VOUCHER:
BRK35